Letts

revise
GCSE

John Hancock
and Alan Bilham-Boult

Geography

Contents

1 Geomorphic processes

2 River landscapes and hydrology

3 Coastal landscapes

4 Glacial landscapes

5 Weather and climate

6 Ecosystems

7 Tectonic activity

8 Population

This book and your GCSE course

	AQA A	AQA B	AQA C	EDEXCEL A
Specification Number	3031	3032	3033	1312
Specification Structure	3 sections all examined	4 regions at different scales, all examined	3 sections, all compulsory	Units A1 - 4 Core Units B5-B6 Options Units C7-C8 Options 1 unit from each option
Terminal Papers	2 x 1 hr 30min. Paper 1: 40% Skills and physical Ans. 3Q from 7 Paper 2: tiered, 35% Human and econ. Ans. 3Q from 6	1 x 1 hr 15 mins Paper 1: 30% UK + OS map Q all Qs compulsory 1 x 2 hr Paper 2: tiered, 45% EU, world & global all Qs compulsory	1 x 1 hr 30 mins Paper 2: tiered, 50% Human, physical, economic 3Qs, all compulsory	1 x 1 hr 45 mins Paper 1: tiered, 45% 4 core units 4 compulsory Qs 1 x 1 hr 15 mins Paper 2: tiered, 30% Option units 2 compulsory Qs on options
Coursework	25%	25%	25%	25%
DME	none	none	25% 1 x 1 hr 30 min all Qs compulsory	none
Physical Geography				
Rocks and weathering (incl. landscape)	9.2	9.1	9.8	
Hydrological cycle and rivers (incl. flooding)	9.4	9.1	9.6, 9.8	1.1, 6.2
Coasts (incl. management)	9.5	12.3	9.8	1.3, 6.1, 7.2
Ice (incl. management of upland glaciated area)	9.3	9.3	9.8	1.2, 7.2
Earthquakes and volcanoes (incl. plate tectonics)	9.1		9.5	5.1, 5.3
Weather and climate (incl. micro-climates and tropical storms)	9.6	9.2, 11.1, 11.2	9.6	4.1, 4.2, 5.1, 5.2
Climate change (incl. global warming, desertification, acid rain, ozone depletion)	9.6	12.3	9.7a, 9.11	6.3, 8.3
Ecosystems (incl. management and sustainability)	9.7	11.1	9.10a	4.3, 6.3
Human and Economic Geography				
Population	10.1	8.4, 11.2, 12.1	9.1, 9.2	2.1, 2.2
Settlement	10.2	9.6	9.3, 9.4	2.3, 8.2, 8.3
Urbanisation	10.2	8.4, 9.6, 10.4, 11.3	9.3, 9.4	8.1, 8.2, 8.3
Energy and resources	10.5	9.4	9.10	3.3
Sustainability	10.5	8.4, 9.1 9.6, 12.3	9.10, 9.11	8.3
Agriculture	10.3	9.2, 10.2, 11.2	9.11	3.1, 3.2
Industry	10.4	9.5, 11.3	9.11	3.1, 3.3
Industry in MEDCs and LEDCs	10.4	11.3	9.11	3.1, 3.3
Industrial change	10.4	9.5	9.11	3.1, 3.3
Development, trade and aid	10.6	8.4, 9.7, 10.5, 12.2	9.9	3.1
Tourism	10.5	9.3, 10.3	9.8, 9.12	7.1, 7.2, 7.3

Visit your awarding body website for full details of your course or download your complete GCSE specification

STAY YOUR COURSE!

Use these pages to get to know your course
- *Make sure you know your exam board*
- *Check which specification you are doing*

- *Know how your course is assessed:*
  - *what format are the papers?*
  - *how is coursework assessed?*
  - *how many papers?*

EDEXCEL B	OCR A	OCR C	WJEC	WJEC (Av.Hill)	NICCEA
1313	1986	1988	3910	1987	
Units A1 - 3 Core Units B4-B5 Options Units C6-C7 Options 1 unit from each option	4 units	5 themes	2 units	4 units	6 themes
1 x 2 hrs Paper 2: tiered, 50% 2 core units (ex DME) 2 compulsory Qs 2 compulsory Qs on chosen options	1 x 2 hr tiered, 50% All units F: 8 compulsory Qs H: 4Qs from 8, 1 from each section 1 x 1 hr tiered, 25% 2 compulsory Qs map interpretation & physical/human links	1 x 2 hr 15 mins, 50% tiered papers All themes Ans. 4Qs from 7: Sect A: Places: 2Q from 3 Sect B: Themes 1 and 2 1Q from 2 Sect C: Themes 3 and 4 1 Q from 2	Paper 1, tiered, 37.5% F: 1 hr 45 mins H: 2 hr Assesses Sect 1 4 compulsory Qs Paper 2, tiered, 37.5% F: 1 hr 45 mins H: 2 hr Assesses Sect 2 4 compulsory Qs	1 x 1 hr 30 mins tiered, 45% Assesses 3 units (ex Paper 2) 3 Qs from 6 1 x 1 hr 30 mins tiered, 30% assesses 1 unit (nominated by exam board in rotation)	2 x 1 hr 30 mins, 80% Paper 1: 40% 3 physical themes 3 compulsory Qs Paper 2: 40% 3 Human/econ. Themes 3 compulsory Qs
25%	25%	20%	25%	25%	20%
25% 1 x 1 hr 15 mins Tiered papers Based on 1 of the core units, all Qs compulsory	none	30% 1 x 1 hr 45 mins Tiered papers Theme 5 all Qs compulsory	none	none	none
7.1 3.4, 4.2, 4.3 3.1, 3.2 7.1, 7.2, 7.3	1a, 1b, 1c 1a. 1b, 1c, 3a 1a, 1b, 1c, 4c	1a 1a, 1c, 2a, 2b, 2c 1a 1a	A1 A4, A5, A6 A4, A5 A2, A3, B5	2.1, 2.2, 2.3, 2.4, 2.5, 2.6 2.4, 2.5, 2.6, 4.9 4.9	B3, B5 B4, B5
3.3	3d	2a, 2b, 2c			B1, B2
4.1, 5.1		1b, 2a, 2b, 2c	B1, B2, B3, B4	1.1, 1.2, 1.3	A1, A2, A4
5.2, 5.3	4a, 4b, 4d	1b	C3	1.3, 1.6, 1.8	A3, C2, C4
	3a, 3b	1c	C2	1.4, 1.5, 1.6, 1.7, 1.8	B5, C1, C2
1.1, 1.2, 1.3, 1.5 2.2, 2.5 2.1, 2.2, 2.3 1.4, 1.6 1.4, 2.3 6.1, 6.2, 6.3 2.4, 2.5 2.5 2.4, 2.5 2.4 7.1, 7.2, 7.3	2a, 2b, 2c 2a, 2b, 2c, 2d 2b, 2c, 2e 3a 3b 3a, 3b 3c, 4a 3c, 4a 3c, 4a 3a, 3b 3d, 4a, 4c	4a 4b, 4c 4d 5a, 5b 5b, 5c 3a, 3b 3a, 3c 3a, 3c 3a, 3c 3c 3b	2C 2A, 2B 2B 1C 1C 2A 2A 2A 2A, 2C 2C 1B	3.6, 3.8 3.3 3.2, 3.4, 3.5, 3.7, 3.8, 3.9 4.7, 4.9 3.9, 4.9 4.1, 4.7 4.1, 4.7 4.8 4.8, 4.9 3.1, 4.2, 4.4, 4.5, 4.6 3.9	D1, D2, D3, D4, D5 F1, F2, F3 F4, F5, F6 D4 D5 E1 E1, E2 E3, E4 E2 E3, E4 E4

Preparing for the examination

Planning your study

The final **three months** before taking your GCSE examination are very important in achieving your best grade. However, the success can be assisted by an organised approach throughout the course.

- After completing a topic in school or college, go through the topic again in your *Revise GCSE Geography Study Guide*. Copy out the main points again on a sheet of paper or use a highlighter pen to emphasise them.
- A couple of days later try to write out these key points from memory. Check differences between what you wrote originally and what you wrote later.
- If you have written your notes on a piece of paper, keep this for revision later.
- Try some questions in the book and check your answers.
- Decide whether you have fully mastered the topic and write down any weaknesses you think you have.

Preparing a revision programme

In the last three months before the final examination go through the list of topics in your Examination Board's specification. Go through and identify which topics you feel you need to concentrate on. It is a temptation at this time to spend valuable revision time on the things you already know and can do. It makes you feel good but does not move you forward.

When you feel you have mastered all the topics, spend time trying past questions. Each time check your answers with the answers given. In the final couple of weeks go back to your summary sheets (or highlighting in the book).

How this book will help you

Revise GCSE Geography Study Guide will help you because:

- It contains the **essential content** for your GCSE course without the extra material that will not be examined.
- It contains **Progress checks** to help you to confirm your understanding.
- It gives **advice from an examiner on how to improve your answers**.
- The summary table will give you a **quick reference** to the requirements for your examination.
- **Marginal comments** and **highlighted key points** will draw to your attention to important things you might otherwise miss.
- It contains important **case studies** to supplement those used in your course and these will help you gain higher marks.

Six ways to improve your grade

1. Planning your time

Make sure you **know how many questions** you have to answer and if there any rules about which ones are compulsory and which include choice. **Plan how much time** you have for each question. Keep to time - do not answer too quickly or too slowly as both can lead to lost marks.

2. Read the question carefully

Many students fail to answer the actual question set. Perhaps they misread the question or answer a similar question they have seen before. **Read the question once right through and then again more slowly**. Some students underline or highlight key words in the question as they read it through. Questions at GCSE contain a lot of information. You should be concerned if you are not using the information in your answer. Command words are usually the first words in a question. They are very important because they tell you exactly what the examiner wants you to do. When you read the question for the first time underline the command word e.g. **describe**, **explain**, **compare**.

3. Give enough detail

If a part of a question is worth **three marks** you should make at least **three separate points** or develop one fully. Be careful that you do not make the same point three times. Approximately 25% of the marks on your final examination papers are awarded for questions requiring longer answers. Practise writing eight to twenty lines of text to answer a question worth 8 to 20 marks.

4. Quality of Written Communication (QWC)

From 2003 some marks on GCSE papers are given for the quality of your written communication. This includes correct sentence structures, correct sequencing of events and use of geographical words.
Read and correct your answer through slowly before moving onto the next part. Remember, you must finish the paper.

5. Using geographical terms correctly

There is an important geographical vocabulary you should use. Try to **use the correct geographical terms** in your answers and spell them correctly. The use of the right terms in the correct place is important in showing your understanding of the topic. Use the right term when labelling diagrams. Using the right term saves time in the examination. As you revise, make a list of geographical terms you meet and check that you understand the meaning of these words.

6. Case studies are important

Many answers gain higher marks when you **use a case study of a real place to illustrate the points you are making**. Make time in your revision to learn them carefully, include local place names and if possible draw a sketch map.

Useful websites

Geomorphic processes

BBC Education – The Essential Guide to Rocks
http://www.bbc.co.uk/education/rocks/ www.bbc.co.uk/education/rocks/

Dartmoor National Park – case study of a granite landscape
"http://www.dartmoor-npa.gov.uk/dnp/homepage.html"
www.dartmoor-npa.gov.uk/dnp/homepage.html

River landscapes and hydrology

Curriculum Visions – an excellent unit on rivers
http://www.curriculumvisions.com/
www.curriculumvisions.com/

Oxfam – case study of flooding in Bangladesh
http://www.oxfam.org.uk/whatnew/features/banglada.htm
www.oxfam.org.uk/whatnew/features/banglada.htm

Coastal landscapes

Geo resources – landforms, erosion and longshore drift
http://www.georesources.co.uk/leld.htm
www.georesources.co.uk/leld.htm

Safeguarding our coast – case study of the Holderness coastline
http://www1.npm.ac.uk/lois/Education/case.htm
www1.npm.ac.uk/lois/Education/case.htm

Glacial landscapes

British Antarctic Survey – research in the Antarctic
http://www.antarctica.ac.uk/ www.antarctica.ac.uk/

Glacier – glaciers and ice sheets in Antarctica
http://glacier.rice.edu/ glacier.rice.edu/

Weather and climate

The Meteorological Office – weather forecasts and education
sections http://www.met-office.gov.uk/ www.met-office.gov.uk/

World climate statistics – 85,000 records of world climate data
http://www.worldclimate.com/ http://www.worldclimate.com/

Ecosystems

The world's biomes – major biomes on Earth
http://www.ucmp.berkeley.edu/glossary/gloss5/biome/index.html www.ucmp.berkeley.edu/glossary/gloss5/biome/index.html

Amazon Interactive – information and activities based on the
Ecuadorian rainforest http://www.eduweb.com/amazon.html
www.eduweb.com/amazon.html

Tectonic activity

United States Geological Survey – information on earthquakes
and volcanoes http://www.usgs.gov/ www.usgs.gov/

Volcano World – everything you could want to know about volcanoes http://volcano.und.nodak.edu/vw.html
volcano.und.nodak.edu/vw.html

Population

US Census Bureau – world population information
http://www.census.gov/ipc/www/world.html
www.census.gov/ipc/www/world.html

Refugees and migration – guide to migration and refugees
http://www.oneworld.org/guides/migration/index.html
www.oneworld.org/guides/migration/index.html

People and the planet – the impacts of population growth on
the Earth http://www.peopleandplanet.net/ www.peopleandplanet.net/

Settlement and urbanisation

Cyber School Bus – interactive site about settlement and urbanisation
http://www.un.org/Pubs/CyberSchoolBus/special/habitat/index.html
www.un.org/Pubs/CyberSchoolBus/special/habitat/index.html

Council for the Protection of Rural England – to protect
http://www.greenchannel.com/cpre/main.htm the countryside
from urban sprawl www.greenchannel.com/cpre/main.htm

NASA pictures of Earth from space –satellite photos and
information on many cities
http://earth.jsc.nasa.gov/ earth.jsc.nasa.gov/

Energy, resources and their management

Energy Quest – useful information on
energyhttp://www.energy.ca.gov/education/
www.energy.ca.gov/education/

Centre for Alternative Technology – alternatives to fossil fuels
http://www.cat.org.uk/information/information.tmpl
www.cat.org.uk/information/information.tmpl

Agriculture

National Farmer's Union – case studies on a large range of different farms http://www.nfu.org.uk/education/default.asp
www.nfu.org.uk/education/default.asp

Friends of the Earth – information on the environmental impacts
of farming http://www.foe.co.uk/ www.foe.co.uk/

Industry

Corus – case study of the steel industry http://www.corus-group.com/ www.corusgroup.com/

Greenpeace – information on the environmental impacts of
industry http://www.greenpeace.org/ www.greenpeace.org/

Development, trade and aid

One World – detailed information on development issues
http://www.oneworld.org/ www.oneworld.org/

The Fairtrade Foundation – trade which ensures fair prices for
the producers http://www.fairtrade.org.uk/
www.fairtrade.org.uk/

Tourism

Virtual Tourist – tourist destinations http://www.vtourist.com/
www.vtourist.com/

Tourism Concern – tourism's impact on people and their environments http://www.tourismconcern.org.uk www.tourismconcern.org.uk

Geomorphic processes

After studying this section you should understand:

- *how to describe a physical landscape*
- *the nature and formation of igneous, sedimentary and metamorphic rocks*
- *the different types of weathering*
- *the role of rocks and weathering in three different landscapes*

Describing the shape of a landscape

A **landscape** consists of a group of **landforms**, which need to be described. Landscapes can be described in the field with reference to a map, and by drawing and labelling a field sketch. A landscape can also be described from a photograph. The following factors should be considered.

- **height** (altitude) in metres, of a number of prominent points or areas
- **relief** – the general shape of the land, the differences in height between the lower and higher areas
- **slope of the land** – steep, gentle, flat
- **shape of the upland areas** – jagged, craggy, smooth
- **shape of the valleys** – flat-floored, U-shaped, V-shaped, gorges
- **nature of the surface of different areas** – rock outcrops, loose rock fragments, soil-covered
- **presence of wate**r – lakes, sea, rivers (meanders), waterfalls
- **vegetation** – moorland, forest, woods, grass, marshes
- **human impact** – quarries, roads, villages, isolated farms, fields
- **personal attitude and feelings** – attractive, awesome, desolate, tedious, exciting.

The examples given for each factor are only examples; there are many more, depending on the landscape being studied.

> Practise reading the contours on a map. Look at the key. What is the contour interval?

> If contours are close together, the slope is steep.

> A V-shaped valley has Y-shaped contours pointing upstream.

Fig 1.1 Annotated field sketch of the Nant Francon Valley in Snowdonia, North Wales as supplied by a GCSE student

Foel-goch (831m)

Minor road just above flat valley floor – only a few isolated farms

Wide, flat-floored valley (210m)

Afon Ogwen – a small river meanders across wide valley floor; wet pastureland

Treeless, heather/grass covered moorland, summer grazing for sheep

Summer grazing for sheep

Steep slopes 200m high with bare rock faces

Steep uneven slopes with many rock outcrops

Narrow river gorge descending towards the main valley ending in a waterfall (Rhaeadr Ogwen Falls)

Braich Ty Du (600m)

West-facing steep slopes (400m) covered in loose rock fragments (scree)

A5T main road, important tourist route into N. Wales

Deciduous woods on lower slopes

Bare rock faces

Thin soils and isolated rocks; tussocky coarse grass and heather

Stunted, isolated trees growing from steep cliffs of rock

Explaining how a landscape has been formed

Understanding and explaining the formation of a landscape, and the variety of landforms that are found there, must take the following factors into account.

- The types of rock.
- The structure of the rocks.
- How different rocks vary in their resistance to:
  - weathering processes
  - the geomorphic processes of erosion, transportation and deposition, resulting from: the work of rivers; the work of the sea; the work of ice; tectonic forces.
- How in the past, when the climate was colder or warmer, weathering and geomorphic processes were different from those affecting the landscape today.

The formation of rocks

AQA A
EDEXCEL B
OCR C
NICCEA

Rocks can be divided into three types based on their formation:

- igneous
- sedimentary
- metamorphic

Igneous rocks

Igneous rocks formed from the molten interior of the Earth. They can be divided into two types.

- **Volcanic** rocks form when molten material, such as lava, is thrown out onto the surface of the Earth and cools quickly.

Fig 1.2 Granite

Feldspar   Quartz   Mica

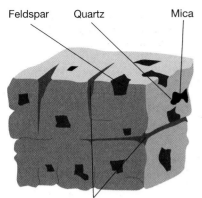

Joints and cracks are the result of pressure release as the rock comes closer to the surface

Understand these terms clearly and use them in your answer.

● **Intrusive** rocks, such as **granite**, form when large masses of molten material are forced up into the crust, and cool slowly under great pressure deep underground.

Some features of granite

● Minerals within the molten rock cool slowly to form large crystals of:

  ● quartz   ● mica   ● feldspar.

● Joints, cracks and weaknesses develop as the rock comes closer to the Earth's surface, and pressure is released.

● Granite is **impermeable**.

● Granite is a hard resistant rock that forms upland areas such as Dartmoor, the Cairngorm Mountains in Scotland (a centre for skiing in the UK) and the dramatic cliffs at Land's End.

● Weathered granite produces **kaolin** used in china, paper and toothpaste.

KEY POINT

An impermeable rock allows very little water to pass through it, most rainwater stays on the surface and runs off.

A permeable rock allows water to:
● **penetrate into the rock through the pore spaces – porous rock**
● **pass along joints and bedding planes – pervious rock.**

Sedimentary rocks

Sedimentary rocks form when rock, which has been eroded by rivers, ice or the sea, is then transported away and deposited. Limestone, chalk, sandstone and clay are all types of sedimentary rock.

Structure of a sedimentary rock

Some examples of sedimentary rock:

● Carboniferous limestone:

  ● a hard grey rock composed of calcium carbonate

  ● has prominent bedding planes and many joints

  ● very **permeable (pervious)** with little surface drainage

  ● forms upland areas such as the **Yorkshire Dales**

  ● used as a building stone, in cement and steel making

● Chalk:

  ● a white or grey rock with many fine pore spaces

  ● has many bedding planes and joints

  ● very **permeable (porous)** with little surface drainage

  ● used in cement making

  ● large amounts of underground water are stored in the pore spaces in the rock called an **aquifer** – important source of water for homes and industry

  ● forms hills such as the **North and South Downs** and the **Chilterns**

Fig 1.3 Sedimentary rock

Deposited in layers or beds of rock of varying thickness

Beds are separated by horizontal bedding planes

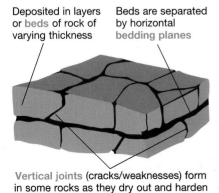

Vertical joints (cracks/weaknesses) form in some rocks as they dry out and harden

- Clay:
  - formed of very fine particles of rock
  - impermeable with lots of surface drainage (marshes, rivers and streams)
  - many river valleys are eroded into clay, e.g. Oxford clay vale from Oxford to Peterborough
  - used for brickmaking, pottery

Metamorphic rocks

Metamorphic rocks form when rocks such as sedimentary rocks are altered by pressure or heat in the crust deep below the Earth's surface. Clay rocks can change to become **slate** as found in north Wales. Limestone can change to become **marble** as found on Iona, north-west Scotland.

Weathering – the processes of rock disintegration

AQA A   AQA C
EDEXCEL B
OCR C
WJEC
NICCEA

 KEY POINT Rocks at or near the surface of the Earth are broken up or decomposed *in situ* by weathering processes.

Rock that has been weathered or broken into fragments is more easily eroded by rivers, ice or the sea and transported away.

There are three types of weathering:

- **physical** (mechanical)   - **chemical**   - **biological**

 Understand clearly the difference between weathering and erosion.

Physical weathering

Freeze-thaw or frost shattering

Fig 1.4 Stages in a freeze-thaw cycle

Rainwater trapped in a joint in a rock

Night Air temperature drops below 0°C, trapped water freezes; water expands in volume as it freezes (by +10%)

1   Water trapped

2   Joint widened

Morning Air temperature rises; ice melts; joint widened by pressure traps more water

Repeated freeze–thaw cycles gradually weaken the joint until the rock splits apart

3

4

Fig 1.5 Cross-section of a scree slope

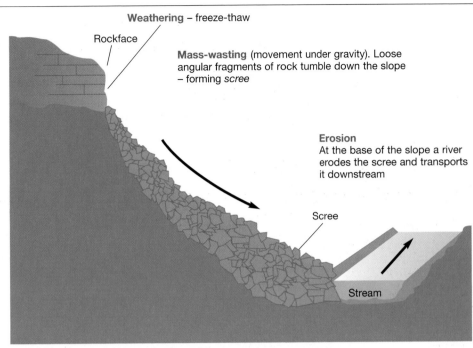

Rockface
– rocks with many joints and bedding planes will trap more water will fragment more easily
– impermeable rocks trap more water
– porous rocks e.g. chalk, store water close to the surface

Weathering
Freezing and thawing more common in
– high mountain areas, e.g. north-west Scotland
– mountain areas in the north of the UK e.g. Lake District
– wetter areas with more rain and snow e.g. north Wales

Weathering – freeze-thaw

Rockface

Mass-wasting (movement under gravity). Loose angular fragments of rock tumble down the slope – forming *scree*

Erosion
At the base of the slope a river erodes the scree and transports it downstream

Scree

Stream

Case study: Nant Francon valley sides (west facing) see Fig 1.1, page 10.

Fig 1.6 Diagram of limestone section

Rain (weak carbonic acid)

Soil

Calcium bi-carbonate carried away in solution

Fig 1.7 Diagram of limestone pavement

clints, upstanding blocks

grykes, deep widened joints

Pressure release

Rocks such as granite cool deep in the Earth's crust, under very high pressures. When exposed at the surface, as on Dartmoor, the pressure has been released and many horizontal and vertical joints and weaknesses have developed in the rock. (See Figs 1.8 and 1.12)

Chemical weathering

Rainwater passing through the atmosphere absorbs carbon dioxide and becomes a weak carbonic acid. The acid reacts with minerals (chemicals) forming the rock, causing decomposition.

Limestone solution (carbonation)

● Limestone is formed of insoluble calcium carbonate.

● Weak acidic water, passing the joints, reacts with calcium carbonate converting it to calcium bi-carbonate, which is soluble and carried away in solution.

● Chemical reactions were faster during periods in the past when climates were warmer than today.

● Joints and bedding planes are slowly widened.

If the soil is removed, a surface is left of upstanding blocks separated by wide joints – limestone pavement.

Granite disintegration (hydrolosis)

● Granite is partly composed of the mineral feldspar.

● Weak acidic water passing along joints and weaknesses reacts with the hydrogen ions in the feldspar, which decomposes into clay (kaolin).

Fig 1.8 Granite disintegration (hydrolysis)

(a) Rain (weak carbonic acid)

(b) A granite tor

- Chemical reactions were faster during periods in the past when climates were warmer than today.

- Loose, disintegrated rock is left surrounding unweathered blocks of granite (Fig. 1.8a).

- If the weathered material is eroded and washed away, a pile of granite blocks will be left – a **tor** (Fig. 1.8b). On Dartmoor, tors are often found on higher points of the moor, e.g. Hounds Tor, Combestone Tor.

Biological weathering

The roots of plants and trees growing in joints and crevices on a rock face expand with age and force the joints apart.

Rotting vegetation creates humic acid, which adds to the processes of chemical weathering.

Landscapes – case studies

AQA A   AQA C
EDEXCEL B
NICCEA

Landforms in limestone areas

Surface landforms caused by solution:

- **limestone pavement** – beds of limestone that have been exposed to give flat areas of rock. Solution has widened the joints called grykes, while the remaining upstanding blocks are known as clints (Fig. 1.7).

- **swallow holes (potholes)** – enlarged joints which allow surface streams to take underground routes which follow bedding planes and joints.

- **limestone gorges** – deep, steep-sided valleys often occupied by a stream, which may be the result of a cavern roof collapsing.

- **limestone scars** – exposed beds of horizontal limestone on valley sides, frequently with areas of scree below them due to freeze-thaw weathering.

- **resurgence** – places where streams re-emerge from an underground course, to flow on the surface again. Often this is at the junction of limestone with impermeable rock below.

Make sure you understand the process of limestone solution (page 13).

PROGRESS CHECK

1. Locate the surface landforms above on the block diagram (Fig. 1.9, page 15).
2. Link their formation to chemical weathering (solution).
3. Answers will benefit from labelled diagrams.
4. Do some research to know located examples.

Underground landforms caused by solution and deposition:

- **caverns and caves** – are the result of widened bedding planes which have often been further enlarged by roof collapse.

- **stalactites and stalagmites** – water seeping and dripping from joints in the roof of a cavern, re-deposits calcium carbonate as features that grow down from the roof ('tites) and up from the floor ('mites).

Case study: Carboniferous limestone (karst) landscape

Fig 1.9 Block diagram of a limestone area such as the Craven district around Malham, Yorkshire Dales National Park (popular tourist areas – dramatic scenery plus visits to underground caverns)

Case study: chalk and clay vale landscapes – South Downs, East Sussex

Fig 1.10 Block diagram of the South Downs and the clay vale (north of Brighton)

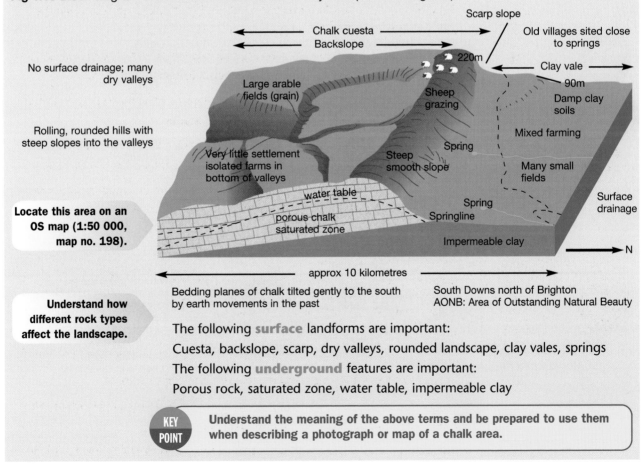

Locate this area on an OS map (1:50 000, map no. 198).

Understand how different rock types affect the landscape.

The following surface landforms are important:

Cuesta, backslope, scarp, dry valleys, rounded landscape, clay vales, springs

The following underground features are important:

Porous rock, saturated zone, water table, impermeable clay

KEY POINT Understand the meaning of the above terms and be prepared to use them when describing a photograph or map of a chalk area.

Case study: granite landscapes – Dartmoor, Devon

Fig 1.11 Annotated map of the Dart valley and surrounding moorland and located tors

Contours drawn at 10m intervals
434 = height in metres
F = isolated farms
Q = disused quarries
◉ = Tor

Low relief on upper moorland

Tors mostly located on tops of hills

Contours close together indicating steep slopes of V-shaped River Dart Valley

A lot of surface drainage

View point and direction of sketch

PROGRESS CHECK

1. Draw a cross-section from Combestone Tor to Yar Tor Down.
2. Label the different parts of the section using ideas from the field sketch.
3. Compare the appearance of this landscape to the limestone area.

Fig 1.12 Field sketch of Combestone Tor and Dartmeet as suplied by a GCSE student

Isolated farms on lower slopes surrounded by small fields

Dartmeet

Woodland in sheltered valley bottom

Combestone granite tor (356m)

Yar Tor (416m)

Treeless open moorland, coarse grass and bracken, summer sheep pastures

River Dart Valley deep V-shaped valley

Grassy slopes eroded by 'tourist path'

Large granite boulders

Juneus grass (reed) indicating wet soils

River landscapes and hydrology

LEARNING SUMMARY

After studying this section, you should understand:

- how a drainage basin functions as a system and is part of the hydrological cycle
- that a storm hydrograph shows changes in discharge following a storm and that the shape of the graph depends on a range of factors
- how rivers erode, transport and deposit material
- that river processes vary between upland and lowland areas
- that there are characteristic landforms associated with upland and lowland river valleys
- the variety of factors that are responsible for flooding
- that flooding is a complex issue to manage

The drainage basin and the hydrological cycle

AQA A
EDEXCEL B
OCR A   OCR C
WJEC   Av. Hill
NICCEA

The **drainage basin** is the area of land drained by a river and its tributaries. It is defined from other basins by its **watershed**.

Fig 2.1 The drainage basin of the River Adur, Sussex

The **gauging station** measures the discharge of the river or the amount of water from the shaded area on the map

Large number of rivers on the impermeable clay **(high drainage density)**

Very few rivers on the porous chalk **(low drainage density)**

The drainage basin forms part of the hydrological cycle.

Fig 2.2 The hydrological cycle (a closed system – no water is gained or lost)

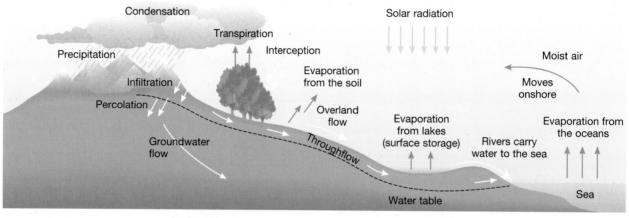

> This is a systems diagram. It is a useful way to remember how the hydrological cycle works.

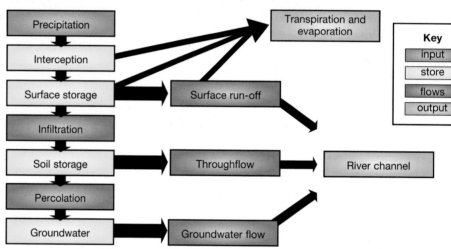

Factors affecting the hydrological cycle

> The hydrological or water cycle is the continuous movement of water between the sea, the atmosphere and the land.

Climate:
- large amounts of rain or snowmelt over a period of time will cause saturation of soils and increased overland flow
- violent storms – increase in overland flow
- hot summer weather and light showers – increased transpiration and evaporation

Vegetation:
- woodland and forest, particularly in summer, intercept more rainfall
  - increased transpiration (loss of moisture from a plant, normally through its leaves)
  - decreased infiltration (movement of moisture from the surface into the soil or rock)

Slopes:
- steep slopes encourage overland flow and decreased infiltration

Soils:
- deep permeable soils increase infiltration and throughflow

Rock type:
- impermeable rocks increase overland flow
- porous rocks increase groundwater storage and flow

River discharge and the hydrograph

The **discharge** of a river is **the amount or volume of water flowing past a particular point at any given time.**

KEY POINT

Discharge (Q) is the average velocity (V) of the river multiplied by the cross-sectional area (A) of the river.
$$Q = V \times A = (m/sec \times m^2/sec) = m^3/sec$$
(cumecs i.e. cubic metres per second)

Fig 2.3 Measuring river discharge

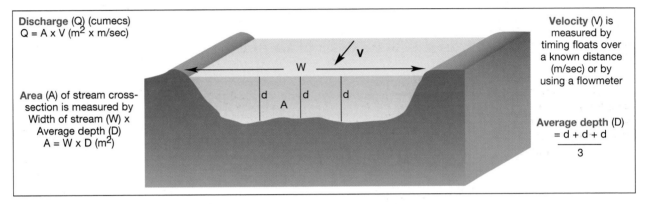

Discharge (Q) (cumecs)
$$Q = A \times V \; (m^2 \times m/sec)$$

Area (A) of stream cross-section is measured by Width of stream (W) x Average depth (D)
$$A = W \times D \; (m^2)$$

Velocity (V) is measured by timing floats over a known distance (m/sec) or by using a flowmeter

Average depth (D)
$$= \frac{d + d + d}{3}$$

Official measurements of discharge are taken at gauging stations along the river. The changes in discharge over a period of time are recorded on a hydrograph. The **flood hydrograph** shows how a river responds to an individual storm.

Fig 2.4 The flood hydrograph

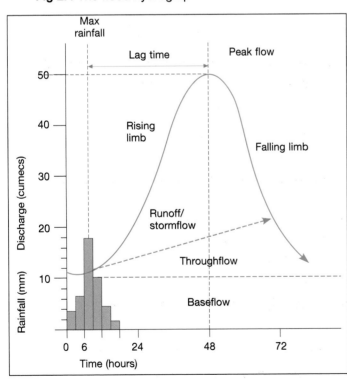

- A storm has been recorded in the drainage basin; during the first 12 hours, maximum rainfall was at 6 hours.

- The rising limb shows discharge slowly increasing as rainfall close to the gauging station finds its way into the river by overland flow.

- Discharge increases more rapidly after 24 hours, as rain which fell throughout the drainage basin flows down the river.

- A peak discharge of 50 cumecs is recorded at 48 hours.

- The lag time is 42 hours (48 – 6 = 42).

 KEY POINT

Lag time is the time difference between maximum rainfall and peak flow. Lag time is important; rivers with a short lag time are more likely to flood.

> Relate these factors to the inputs, stores and flows in Fig. 2.2.

Factors affecting lag time and peak flow

Climate: rainfall; snowmelt

River basin: vegetation; slopes; soil; rock type

Human activity can decrease lag time by: draining more fields by underground drainage pipes; removing woodland and forests (deforestation); building more houses and roads (urbanisation); straightening river channels and artificially raising the height of river banks.

Human activity can increase lag time by: planting more woodlands and forest; building dams to store and control discharge; extracting water for industrial and domestic use; allowing flood water to be temporarily stored on the floodplain or in 'washes' during the winter.

A decrease in lag time can increase the likelihood of flooding.

> This often appears in exam questions. Make sure you can explain why they will affect lag time. Use the correct terms from Fig. 2.2.

River processes and landforms

AQA A   AQA B   AQA C
EDEXCEL A
OCR A   OCR C
WJEC   Av. Hill
NICCEA

Rivers flow from their source to the sea. The slope the river follows is called the **long profile**.

Fig 2.5 The long profile of a river

> River energy – energy is required for rivers to erode and transport material. Loss of energy causes rivers to deposit material.

River source

Sea

Upland valley
- slope is steep and uneven (many rapids and waterfalls)

Lowland valley
- slope is gentle and smooth

- discharge increases downstream as more tributaries join main river ⟶

- water requires **energy** to flow downhill
- 95% of energy used to overcome friction and turbulence (t)

2m wide bankfull stage   0.5m

Velocity increases

Energy increases

10m wide bankfull stage   2.3m

- small, rocky channels
- lots of friction with the rough channel
- very turbulent water
- high energy loss

- wide, deep channels with a smooth shape
- less friction
- less turbulence
- much energy to erode and transport

- in upland and lowland valleys rivers flow fastest when the channel is full – **bankfull stage**

- when discharge is high, there will be maximum energy to overcome friction and turbulence

- at bankfull stage there will be maximum erosion and transportation

River processes

Erosion – wearing away the banks and bed of the river channel.

Transportation – the movement of eroded sediments (boulders, sand and mud) downstream.

Deposition – transported sediments are deposited by the river.

- Erosion and transportation are closely linked since the transported sediments greatly increase the ability of the river to erode.
- The amount of erosion and transportation is greatest when the river is at bankfull stage and there is most surplus energy.

Erosion – takes place by four processes:

> **Learn and use these terms correctly.**

- **abrasion or corrasion** – sediments carried along rub and abrade the channel sides and bed; pebbles swirled in cavities in the bed drill downwards forming circular **potholes**

- **attrition** – transported rocks collide, break up and become smaller and more rounded downstream

- **hydraulic action** – the immense weight and force of the flowing water removes material from the bed and sides of the channel

- **corrosion** – water reacts with minerals (chemicals) in the rocks which are carried away in solution. This is important where rivers pass over areas of limestone and chalk.

Transportation – sediments are transported in four ways:

- traction
- suspension
- saltation
- solution

> **KEY POINT** A river's load is the amount of sediment being transported.

Fig 2.6 Processes of transportation

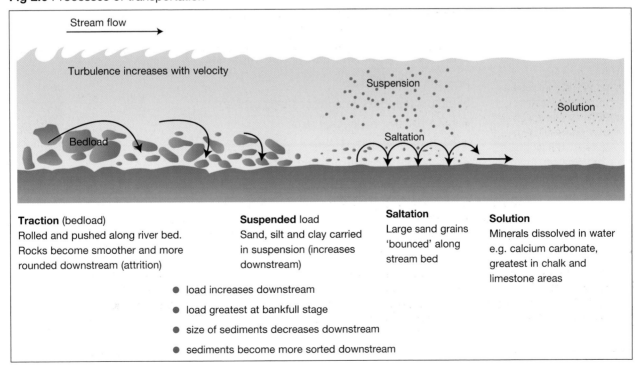

Traction (bedload)
Rolled and pushed along river bed. Rocks become smoother and more rounded downstream (attrition)

Suspended load
Sand, silt and clay carried in suspension (increases downstream)

Saltation
Large sand grains 'bounced' along stream bed

Solution
Minerals dissolved in water e.g. calcium carbonate, greatest in chalk and limestone areas

- load increases downstream
- load greatest at bankfull stage
- size of sediments decreases downstream
- sediments become more sorted downstream

Deposition – occurs when a river lacks sufficient energy to transport the load it is carrying. Often this is when the river loses velocity:

- on entering a lake or the sea
- on the inside of a bend or meander.

Fig 2.7 Rivers deposit their load on entering a lake

River landforms

AQA A AQA B AQA C
EDEXCEL A
OCR A OCR C
WJEC Av. Hill
NICCEA

River landforms in an upland area

Many rivers have their source and headwaters in upland areas; the Rivers Severn and Dee rise in upland Wales.

Upland areas in the UK have:

- high precipitation with heavy snowfalls
- steep slopes
- impermeable rock
- thin soils and sparse vegetation
- areas where overland flow is very important.

Rivers in upland areas:

- have steep, uneven courses (long profile)
- have narrow, rocky channels
- have short lag times
- have rapid increases in discharge
- are very turbulent
- can transport boulders, rolling them along the bed by traction
- mainly erode downwards – *vertical erosion*.

Link these characteristics together to come to a better understanding of how the hydrological cycle works in upland areas.

Questions frequently ask for labelled diagrams explaining the formation of river landforms.

V-shaped valleys and interlocking spurs

E.g. the River Dart valley, Dartmoor (see page 16).

Fig 2.8 The formation of V-shaped valleys and interlocking spurs

V-shaped valley
Cross-section of an upland valley

River erodes downwards by abrasion/corrasion

- as the valley deepens the sides become too steep and unstable
- rock tumbles and slides into the valley bottom (mass wasting)
- loose rock is carried away by the stream
- the valley widens to a V-shape

Interlocking spurs
(view down valley)

Stream flowing between interlocking spurs of hard rock

Stream

> Remember to use and name real examples you have seen on field trips or holidays.

Waterfalls and gorges

Where rivers cross beds of harder (resistant) and softer rocks, waterfalls may form together with deep, steep-sided gorges, e.g. High Force on the River Tees, in the north Pennines.

Fig 2.9 Formation of waterfalls and gorges

Overhanging rock eventually collapses – waterfall moves upstream

Steep-sided gorge left as waterfall retreats upstream

Hard rock

Softer rock undercut

Very turbulent water

> Use process terms to make your explanations clearer.

Hydraulic pressure – great turbulence causes corrasion and creates a deep plunge pool

Rocks in the plunge pool swirled around and erode it deeper (abrasion)

Waterfalls may also be the result of:

- glacial erosion (see page 45)
- a fall in sea level.
- Earth movements

River landforms in a lowland area

Rivers such as the Severn and Dee flow through wide, lowland valleys before reaching the sea.

Lowland areas in the UK, compared with the uplands, have:

- lower precipitation
- gentler slopes
- less resistant rocks
- deeper soils and more vegetation
- areas where throughflow, groundwater flow and storage are more important than overland flow.

Rivers in lowland areas:
- have low gradients, gentle long profiles
- have wide, deep, smooth channels
- have longer lag times
- have larger discharges (from the whole drainage basin) particularly at bankfull stage
- have stronger currents (velocity)
- have lower friction and turbulence
- have increased erosion and transportation
- have erosion which is mostly side-to-side – lateral erosion
- transport large 'loads' of fine sediment by saltation and suspension.

Meanders, floodplains and oxbow lakes

With large discharges, strong currents (velocity), low friction and less turbulence, rivers in lowland areas have surplus energy to erode and transport. Lateral erosion erodes the banks of the channel and large bends or meanders develop. Since erosion occurs mostly on the outside of bends, where velocity is greatest, meanders increase in width. The valley sides are slowly eroded to create wide, flat floodplains. Meanders can develop extreme loops across the flooplain. Further erosion can leave 'loops' abandoned as oxbow lakes, when the river assumes a straighter, more efficient course.

Fig 2.10 Formation of meanders, flood plains and oxbow lakes

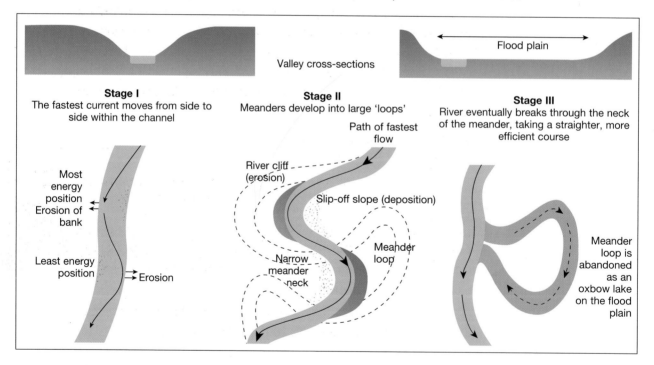

Fig 2.11 Map of River Dee showing floodplain, meanders and oxbow lakes (based on OS map)

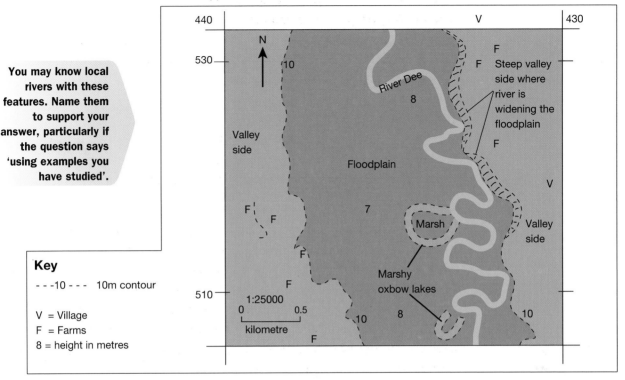

> You may know local rivers with these features. Name them to support your answer, particularly if the question says 'using examples you have studied'.

Key

- - -10 - - - 10m contour

V = Village
F = Farms
8 = height in metres

At times of very high discharge, rivers may overflow their banks and spread out across the floodplain. Floodplains are natural 'stores' of water until the floodwaters go down. At bankfull stage the river carries maximum 'load'. On flooding, the water slows suddenly, energy is lost, and deposition takes place. Coarse sediments (sand) are deposited immediately on the river banks, to form natural **levées**. Finer sediment (silt and clay) forms deposits across the floodplain called **alluvium**.

Fig 2.12 Floodplains and levées

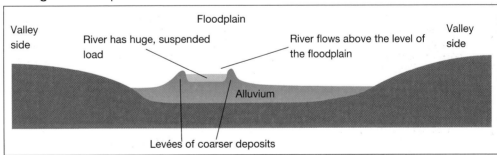

Alluvium is very fertile. In the lower Nile valley in Egypt, farmers relied on annual floods to bring fresh supplies of silt to their land.

Deltas

Deltas form at the mouths of many of the world's larger rivers, e.g. the Nile (Egypt), the Ganges (Bangladesh), the Mississippi (USA) and the Mekong (Vietnam).

Deltas form where:

- rivers carry huge 'loads' of fine sediments
- currents slow on entering the sea causing deposition
- tidal currents are not strong enough to remove the sediment
- small deltas can form where rivers enter lakes (Fig. 2.7).

> Suggest reasons why, in many LEDCs, there are dense populations of people living on deltas. Why might these be dangerous places to live?

The river is broken up into many channels, called distributaries, which spread the sediment out into distinctive shapes.

Types of delta:

Fig 2.13 Types of delta

- arcuate, e.g. Mekong and Nile
- bird's foot, e.g. Mississippi

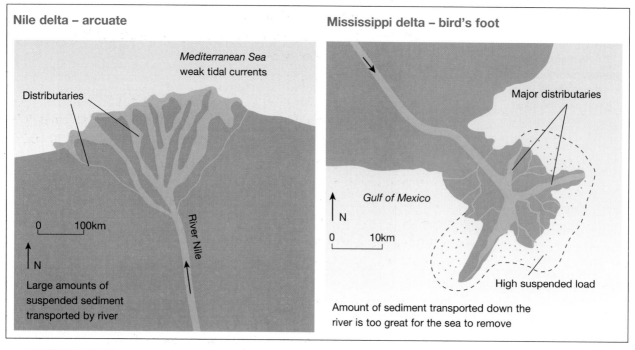

Nile delta – arcuate

Mediterranean Sea
weak tidal currents

Distributaries

0   100km

N

River Nile

Large amounts of suspended sediment transported by river

Mississippi delta – bird's foot

Major distributaries

Gulf of Mexico

N

0   10km

High suspended load

Amount of sediment transported down the river is too great for the sea to remove

Estuaries

Questions often link estuaries to the location of industries such as oil refineries. Why are they located there?

There are no large deltas around the coasts of the UK. Many rivers have estuaries, which are wide river valleys flooded daily by the tide. At low tide, large expanses of mud and sandflats are exposed which are mostly composed of material deposited by the river, e.g. Thames, Humber, Severn.

River flooding and management

AQA A   AQA B   AQA C
EDEXCEL A   EDEXCEL B
EDEXCEL C
OCR A   OCR C
WJEC   Av. Hill
NICCEA

KEY POINT Flooding occurs when the discharge of a river is too great for its channel to hold. Water will flow over the banks and occupy the floodplain.

Floods can have huge social and economic costs, but they can also be beneficial. Vast numbers of people around the world live in areas that suffer from regular flooding.

Causes of flooding

Most flooding is a combination of climatic, drainage basin and human factors.

Climate factors:	• heavy rainfall is less likely to infiltrate into the soil and will reach the river quickly
	• rapid snowmelt contributed to the 1995 Rhine floods
Drainage basin factors:	• impermeable rock, thin soils, steep slopes and saturated ground
Human activity factors:	• deforestation reduces interception and transpiration
	• urbanisation and modern farming practices increase run-off

Impacts of flooding

- loss of life – people evacuated from their homes
- buildings and property damaged – high insurance costs
- crops ruined, animals drowned
- transport and communications flooded
- economic impact on shops, offices and industry, and public utilities
- sewage contaminates water supplies
- political pressures on government

Impacts in LEDCs are likely to be more severe than in MEDCs.

- Governments cannot afford to invest in costly flood prevention schemes. It is estimated that schemes in Bangladesh will cost over £1 billion over 30 years (Bangladesh already has a large international debt).
- Governments give higher priority to investing in industry and exports, to raise their GNP, than in flood prevention schemes.
- Schemes funded by the World Bank provide only long-term solutions – aid agencies are only able to support schemes to improve farming, schools, hospitals which will improve survival rates rather than schemes to prevent flooding.
- Dense populations are attracted to the fertile soils on floodplains and deltas where subsistence agriculture (see page 147) provides them with food (e.g. Mekong Delta) – these areas are likely to suffer annual floods.
- Living at subsistence level, a flood can destroy crops and livestock, causing food shortages, starvation and disease.
- Poor communications and medical services mean help is slow to arrive.
- Flood warning systems not well developed, many people are without radio and television.

Flood control

- **Dams** can store and control river discharge (dams are used to control tributaries of the Mississippi, e.g. River Missouri).
- **Levées and retaining walls** – by raising the height of river banks water can be contained. (November 2000 – flood walls in York contained record river levels and saved many homes from flooding.)
- **Straightening meanders** – increases speed of flow and reduces the length of the river (the length of the Mississippi has been shortened by over 240km).
- **Flood relief channels** – provide additional channel alongside existing course (the Maidenhead, Windsor and Eton Relief Channel, opening 2001, will save 5500 homes from Thames floodwater).
- **Washlands or spillways** – in times of high discharge sluice-gates are opened and water is allowed to flood adjacent areas (washlands have been in use on the Great Ouse in Cambridgeshire for many centuries).
- **Afforestation** (the planting of woods and forests) – increases interception and reduces run-off (e.g. Tennessee river valley, a tributary of the Mississippi).
- **Planning regulations** – reject development proposals on flood plains so that they can still be allowed to flood (between 1970 and 1990 in the UK, two million homes were built on floodplains).

Failure of dams or levées can cause very severe flooding.

Where rivers cross international boundaries, river management and flood control are more difficult.

● Flood warning systems – In September 2000, the Environment Agency (UK) issued new flood warning codes to be broadcast on the radio and television, and produced flood plan checklists.

MEDC case study: Mississippi floods, August 1993

Causes

● There had been continuous rainfall from April to July; soils were saturated with rapid run-off.

● Heavy thunderstorms in June throughout the upper drainage basins of the Missouri and Mississippi gave over twice the normal rainfall.

● Huge drainage basins (30% of the USA); large rivers meeting at confluences, e.g. St. Louis, combined to give record discharges.

● Silt had raised the bed of the channel; levées had been raised. The river flowed above the level of the surrounding floodplain.

● Record discharges caused levées to collapse and the river eroded huge gaps in the banks and flooded the low-lying floodplains.

● Growth of large urban populations demanded higher flood defences and caused rapid run-off.

> Questions frequently ask for case studies in an MEDC or LEDC. Make sure you can relate your answers to the place you have named.

Fig 2.14 The drainage basin of the Mississippi

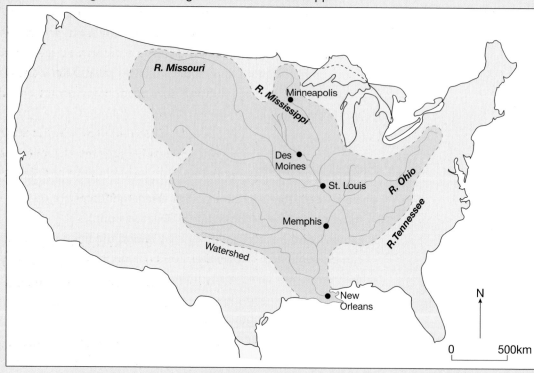

Impacts

● Severe flooding occurred from Minneapolis to Memphis (8 million hectares).

● 47 people died, 74 000 were evacuated and 45 000 homes were affected.

● Des Moines and St. Louis lost electricity and water supplies.

● Sewage contaminated water supplies over large areas.

● Crops of grain and soya bean, valued at $6.5 billion, were ruined.

● Communications disrupted.

● Total cost estimated at $11 billion.

Outcomes

- Government declared a state of emergency; military aid supplemented local emergency services; states received large amounts of federal aid.
- 'Hard' defences soon repaired but more spillways or washes have been designated, to store surplus water.
- Most people had insurance policies to meet the costs of destruction.

LEDC case study: Mekong delta floods, 2000

Fig 2.15 The drainage basin of the Mekong river

Watershed of Mekong drainage basin

HEP dams

Record flood levels covering large areas of the delta and extending almost to the capital city, Ho Chi Minh.

Causes

- The mountainous, steep-sided valleys in Laos and Cambodia received heavy rainfall.
- The monsoon rainfall (wet and dry seasons) was above average for the previous three years, so the ground was very saturated from flooding.
- Much of the rainfall fell in very intense storms.
- Rural communities in Laos and Cambodia depend on the river and forest for building materials, food and medicines; many are shifting cultivators. Population growth is causing deforestation.
- Poverty leads to illegal logging – 39 000 recorded violations of forestry regulations in one year.
- Concrete-lined riverbanks increasing in urban areas such as Vientiane, the capital of Laos.
- Dam building in Laos and China causing deforestation.
- Dams in China and Laos released large discharges of water without regard for the delta areas in Vietnam.
- Dense rural population living in the delta causing overgrazing and soil erosion.

Impacts

- 350 people dead; mostly young children.
- Many hundreds of thousands forced to leave their homes in June to occupy small muddy banks. They were unable to return to their villages until December.
- Three successive years of flooding left no reserve food supplies; people had to rely entirely on relief workers to bring food.
- New high yielding rice strains are unable to withstand long floods; the year's rice crop was lost.
- A plague of golden water snails threatened all surviving crops.
- There were life-threatening diseases from contaminated water.

> Make a simple table to compare impacts and outcomes with an MEDC case study.

Outcomes

- The government has extended small loans so that families can build houses on stilts. It has also distributed thousands of small boats so that families can continue to fish for food during floods.
- The government has implemented a successful policy restricting families to two children only; population growth reduced.
- Dams in Laos will help control floods but will also reduce the amount of silt carried down by the river. Delta farmers rely on fresh fertile silt to replenish their land. The lack of silt will cause the delta to shrink in size and become lower – threatening worse flooding.

PROGRESS CHECK

1. What is the area drained by a river and its tributaries called?
2. What is the flow of water downhill, just below the surface, called?
3. What are the units of discharge measured in?
4. On a hydrograph, what is the name given to the time difference between maximum rainfall and peak flow?
5. What is the name given to the movement of sediment along a streambed by short hops or jumps?
6. What shape would suggest to you that an upland valley had been eroded by a river?
7. What is the name given to an abandoned 'loop' of a river found on a floodplain?
8. What name is given to wide river valleys flooded daily by the tide such as the Humber?

1. Drainage basin 2. Throughflow 3. Cumecs (c^3m/sec) 4. Lag time 5. Saltation
6. V-shape 7. Oxbow 8. Estuaries

3 Coastal landscapes

LEARNING SUMMARY

After studying this section you should understand:

- how waves form and the different types of wave
- the processes of erosion, transportation and deposition
- the landforms associated with erosion and deposition
- how coastal defences can be managed to prevent flooding
- that tourism and other activities along a coast have to be managed

KEY POINT   The coast is the interface or narrow zone where the sea, land and atmosphere meet.

Coasts are shaped by the processes of erosion, transportation and deposition. The forces most responsible for these processes are:

- **waves**   - **tides**   - **currents.**

Coasts can change shape rapidly, as in a single storm, but most landforms are the result of long-term changes. The nature of the rocks forming the coast affects its shape and form.

Formation of waves

- Waves result from the wind blowing over the surface of the sea. (This is not the case for tsunamis; see page 101.)
- The largest waves have the most energy.
- Wave size depends on:
  - the strength of the wind - the duration of the wind   - the fetch.

Fig 3.1 The formation of waves

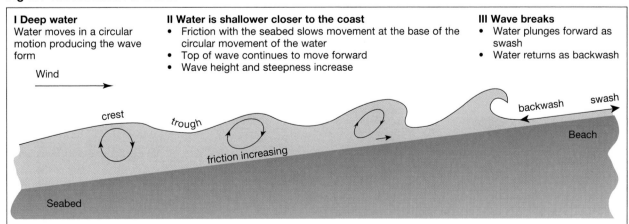

I Deep water	II Water is shallower closer to the coast	III Wave breaks
Water moves in a circular motion producing the wave form	• Friction with the seabed slows movement at the base of the circular movement of the water • Top of wave continues to move forward • Wave height and steepness increase	• Water plunges forward as swash • Water returns as backwash

Wind

crest   trough   friction increasing   backwash   swash

Beach

Seabed

KEY POINT The **fetch** is the distance the wind has travelled over the sea before reaching the coast.

At Land's End, Cornwall, waves with the highest energy come from the south-west (Atlantic Ocean) because:

- frequent storms over the Atlantic give gale force winds
- winds blowing from the south-west are the **prevailing winds**
- winds from the south-west have a fetch of many thousands of miles.

Fig 3.2 Types of waves

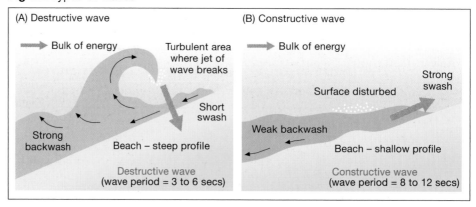

Tides

At high tide the water will be deepest offshore and larger waves with more energy can reach the beach or cliff. Storm waves at high tide have the highest energy of all and they are responsible for most erosion and transportation.

Processes of erosion

 AQA A   AQA C
EDEXCEL A   EDEXCEL B
OCR A   OCR C
WJEC   Av. Hill

Waves can erode in four different ways:

- **hydraulic action:** the pressure of the wave hitting a cliff traps and compresses air into joints and cracks, weakening the rock and breaking it up
- **abrasion (corrasion):** waves hurl rocks against the cliff and swirl sand backwards and forwards, wearing the rock away
- **attrition:** rocks and sand rub against each other, making them smaller and more rounded
- **corrosion (solution):** salt water corrodes minerals in many types of rock, causing rocks to break up.

 Note the similar terms to rivers.

Weathering processes (see page 12) such as freeze-thaw break up the rock on the cliff face.

Mass movement: cliffs collapse when the waves erode and undercut the base of the cliff.

Landforms of erosion

AQA A AQA C
EDEXCEL A EDEXCEL B
OCR A OCR C
WJEC Av. Hill

Headlands and bays

Fig 3.3 Formation of headlands and bays

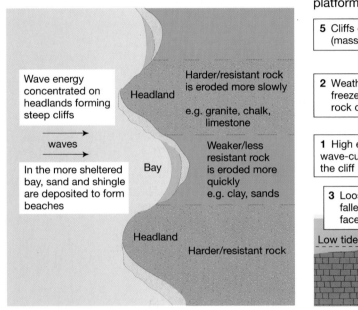

Wave energy concentrated on headlands forming steep cliffs

Headland

Harder/resistant rock is eroded more slowly

e.g. granite, chalk, limestone

waves

In the more sheltered bay, sand and shingle are deposited to form beaches

Bay

Weaker/less resistant rock is eroded more quickly e.g. clay, sands

Headland

Harder/resistant rock

> Link to actual examples you have studied.

Cliffs and wave-cut platforms

Fig 3.4 Stages in the formation of cliffs and wave-cut platforms

5 Cliffs eventually collapse as they are undercut (mass movement) and retreat

2 Weathering (e.g. freeze-thaw) breaks up the rock on the cliff face

1 High energy waves erode a wave-cut notch, undercutting the cliff

3 Loose rocks fallen from cliff face – corrasion

4 Wave-cut notch – erosion along bedding planes and joints

High tide

Low tide

6 Wave cut platform – a gently sloping rock surface exposed at low tide extends as cliff is eroded (retreats)

Caves, arches, stacks and stumps

Fig 3.5 Formation of caves, arches, stacks and stumps

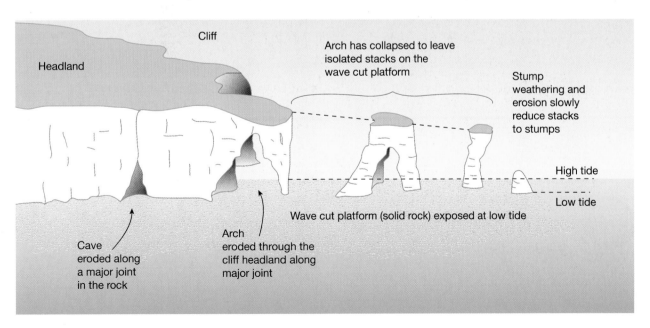

Cliff

Headland

Arch has collapsed to leave isolated stacks on the wave cut platform

Stump weathering and erosion slowly reduce stacks to stumps

High tide

Low tide

Cave eroded along a major joint in the rock

Arch eroded through the cliff headland along major joint

Wave cut platform (solid rock) exposed at low tide

Case study: Coastal landforms, geology and processes along the Dorset coast

Fig 3.6 Coastal landforms, geology and processes along the Dorset coast

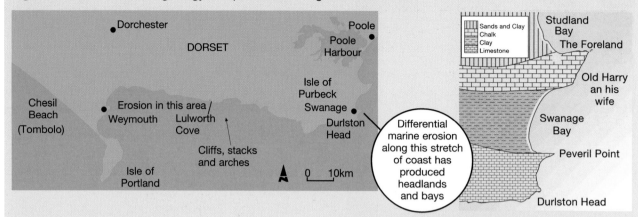

- Pronounced headlands of limestone and chalk
- Wide bays where the softer clays have been more easily eroded

- Sandy beaches are found in the shallow and sheltered waters of the bays
- Stacks and stumps are found off the headlands (Old Harry)

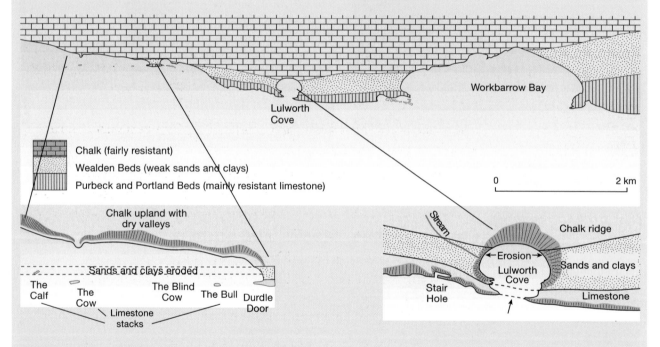

- Limestone has almost gone – a few isolated stacks and stumps remain offshore (The Bull)
- Durdle Door is an arch eroded through the narrow limestone

- At Lulworth Cove the sea has broken through the limestone
- The soft sands and clays are more easily eroded
- The chalk is more resistant and forms a steep cliff at the back of the cove

---

PROGRESS CHECK

Draw a labelled diagram to show how Workbarrow Bay has been formed.

Processes of transportation

Waves transport:

- rocks and sediments eroded from the cliffs
- sediments carried down to the sea by rivers.

Sediments are moved up and down the beach (swash and backwash).

Sediments are moved along the beach by **longshore drift**.

Fig 3.7 Longshore drift

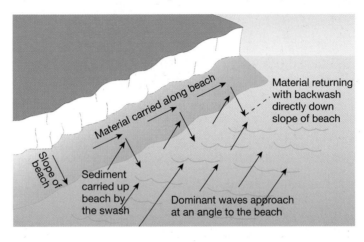

Landforms of deposition

Beaches

Beaches are the most common landforms resulting from deposition. Beaches are formed of sand and shingle. Beaches form where:

- sufficient sediment is available from cliff erosion or offshore deposits
- longshore drift maintains a constant movement of sediment along the coast
- waves lack sufficient energy to transport sediment, because the water is too shallow or sheltered, such as in a bay.

The size, shape and height of a beach depends on:

- the nature of the beach material
- the dominant type of wave

KEY POINT **Tidal range is the difference in the height of the sea between high tide and low tide.**

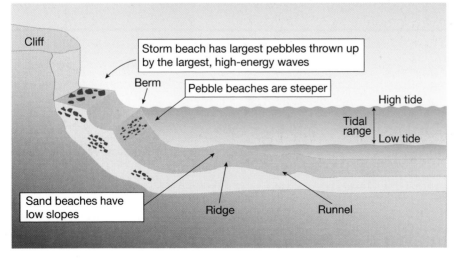

Beaches absorb wave energy. High, wide beaches:

- protect cliffs from erosion by the waves
- are the best form of coastal defence.

Fig 3.8 Beaches differ in shape, height and size

Coastal landscapes

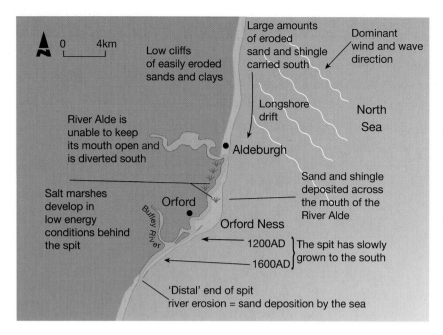

Fig 3.9 Spit formation at Orford Ness, Suffolk

Fig 3.10 Chesil Beach, Dorset

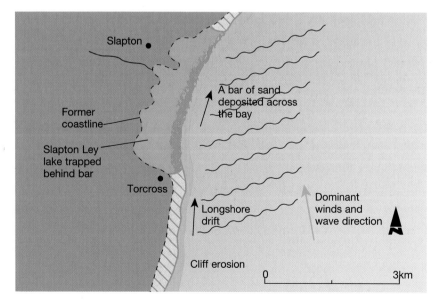

Fig 3.11 Sand bar at Slapton Ley, Devon

Spits

Spits are:

- formed of sand and shingle
- the result of longshore drift
- attached to the mainland at one end
- frequently found where coasts change direction at the mouths of rivers or estuaries.

Tombolos

Tombolos are similar to spits, but link the mainland to an offshore island as at Chesil Beach in Dorset.

- south-west prevailing winds
- very long fetch
- high energy waves

Bars

Bars are similar to spits, but they form right across a bay, from headland to headland.

Coastal management

AQA A · AQA C · EDEXCEL A · EDEXCEL B · OCR A · OCR C · WJEC · Av. Hill

Coasts need to be **managed** to:

- sustain human activities from the threat of coastal erosion
- preserve coasts for heritage and conservation reasons
- preserve coasts from over development.

Managing coastal defences

Coastal defence is necessary in the UK because:

- over 17 million people live within 10km of the coast
- 40% of manufacturing industry is on or near the coast
- 31% of the coastline has investments in roads, buildings and recreation facilities
- the pressure to develop coastal sites for housing and industry, including tourism and leisure, is likely to increase.

Responsibility lies mainly with:

- **The Environment Agency** (national government agency – all coasts)
- **MAFF** (national government agency – all coasts)
- **district councils** (local coasts).

To obtain planning permission, schemes must satisfy three basic criteria:

- engineers consider the scheme to be viable
- it is economically viable (cost-benefit analysis)
- it is environmentally sound.

KEY POINT

Cost-benefit analysis assesses whether the benefits gained from having a sea defence (e.g. the costs of flooding, loss of life etc.) outweigh the costs of construction.

Planning options are shown in Table 3.1 below.

Table 3.1 Planning options

Do nothing	Prevent and discourage	Managed retreat	Build defences
Allow natural processes of erosion to continue, until a new balance has been achieved. Unlikely option where urban areas need defending.	Planning controls prevent further development at vulnerable sites. Mortgages refused and cost of insurance increased.	Defend the coast further inland and allow the present coast to be eroded. Provide compensation and resettlement schemes for farmers and house owners etc. Viable only in areas of low population?	'Hard' defences such as sea walls, rip-rap, groynes. 'Soft' defences such as beach nourishment.

Use examples from a coast you have studied

Types of coastal defence are shown in Table 3.2 on page 38.

Table 3.2 Types of coastal defence

Description	Effects	Advantages	Disadvantages
Type: Seawall			
Concrete wall of different designs; can include flat 'decks' for walking or car parking. £3000–4000/m	Deflects wave energy back into the next wave.	Capable of withstanding high energy waves. Efficient and effective. Strength and height reassures public fear of flooding.	High cost of building and repair. Reflection can cause turbulence and scouring of the beach. Not very attractive. Access to beach difficult.
Type: Rip-rap (rock armour)			
Jumble barrier of large, irregularly shaped rocks or concrete. £3000/m	Absorbs wave energy in gaps and voids between the boulders.	Very efficient. Relatively cheap.	Costs rise if rock imported (e.g. Norway). Rocks must be large enough to remain stable (3 tonnes). Difficult access to beach; dangerous for young children.
Type: Gabions			
Strong wire baskets filled with stones and rubble. Often used to protect sand dunes. £100/m	Gaps between stones absorb wave energy.	Trap sand, often become covered with vegetation. Visually less obtrusive.	Not as effective as a sea wall. Shorter life span if exposed.
Type: Groynes			
Low walls of timber or concrete built at right angles across the beach. £7000 each	Traps sand moving along the beach by longshore drift. Beach built up between the groynes is able to absorb more wave energy.	Cheaper than sea walls. Maintains beach for tourists. No problems of access	Beaches further along the coast maybe starved of a supply of sand and become more vulnerable.
Type: Revetment			
Low sloping walls of concrete built at the top of the beach. £2000/m	Reflects and absorbs wave energy.	Beach scouring is less than with a sea wall. Relatively cheap form of 'hard' defence. Less intrusive than a sea wall.	Unsuitable for high energy conditions. Short life span.
Type: Embankment			
An earth bank dug from nearby and covered with grass. Often found along banks of tidal estuaries and rivers with a saltmarsh in front.	Prevents water at high tide flooding the low-lying land behind. Only found along coasts where wave energy is low.	Relatively low cost. Cheap maintenance.	Unable to withstand high energy waves. Erodes quickly if water overflows the bank.
Type: Beach nourishment ('soft' defence)			
Beach made higher and wider by feeding sand and shingle brought in by lorries or dredgers. Often used to 'protect' hard defences. £20/cu.m	Beach more able to absorb wave energy, particularly in storm conditions.	Very effective. Relatively cheap. Maintains natural appearance of coast. Preserves a beach for leisure purposes.	Beaches probably need to be re-nourished if storms cause erosion, increasing the cost. Offshore dredging may increase erosion in another location.

Fig 3.12 Flood risk and beach nourishment sites along the east coast

- Dredging site
- Beach nourishment sites
- Flood risk areas

0 ————— 50km

Skegness, Kings Lynn, Norwich, Great Yarmouth, Peterborough, Ipswich, Harwich

Case study: Holderness and Lincolnshire coasts

The east coasts of the UK south of Flamborough Head (Yorkshire) are vulnerable to erosion and flooding because:

- much of the area is low lying with many parts less than 5m above average sea-level
- cliffs are formed of soft, easily-eroded rocks such as glacial till (boulder clay)
- East Anglia is slowly sinking relative to sea-level
- global warming is likely to cause:
  - a rise in sea-level, particularly threatening low-lying coasts
  - increased storminess in the North Sea and more storm surges.

Fig 3.13 Management problems and solution along the Holderness and Lincolnshire coasts

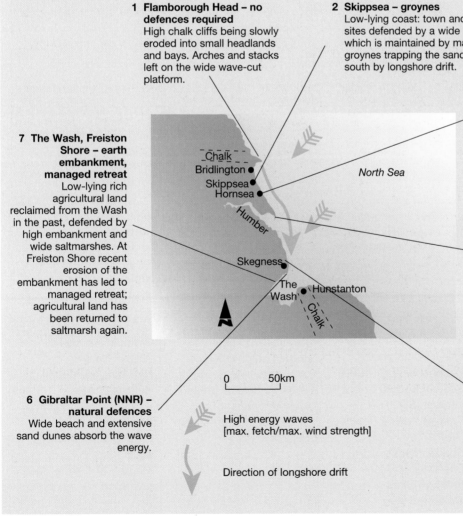

1 Flamborough Head – no defences required
High chalk cliffs being slowly eroded into small headlands and bays. Arches and stacks left on the wide wave-cut platform.

2 Skippsea – groynes
Low-lying coast: town and caravan sites defended by a wide beach which is maintained by many wooden groynes trapping the sand moved south by longshore drift.

3 Hornsea – high seawall and groynes
After they were constructed in 1924, beaches *to the south* were starved of sand. The cliffs in soft glacial till (boulder clay) were eroded much more quickly. Erosion increased from 1.9m to 4.5m/year with the loss of farmland, houses and roads.

4 Spurn Head – rip-rap
Erosion has almost broken through the spit at its northern end; rip-rap has been used to absorb the wave energy. There is a danger of the road which leads to the lifeboat station and Humber pilots' base at the end of the spit being cut off.

5 Mablethorpe to Skegness – heavily defended by seawalls and beach nourishment
Low-lying land, much of it is below high tide. The towns are growing with new estates of bungalows mostly for retired people, and there are large investments in tourist industry. Defences are expensive to maintain.

6 Gibraltar Point (NNR) – natural defences
Wide beach and extensive sand dunes absorb the wave energy.

7 The Wash, Freiston Shore – earth embankment, managed retreat
Low-lying rich agricultural land reclaimed from the Wash in the past, defended by high embankment and wide saltmarshes. At Freiston Shore recent erosion of the embankment has led to managed retreat; agricultural land has been returned to saltmarsh again.

Chalk, Bridlington, Skippsea, Hornsea, Humber, North Sea, Skegness, The Wash, Hunstanton, Chalk

0 ————— 50km

High energy waves [max. fetch/max. wind strength]

Direction of longshore drift

Managing coasts for heritage and conservation

Case study: Dorset Heritage Coast

Fig 3.14 Conflicting demands on a coastline?

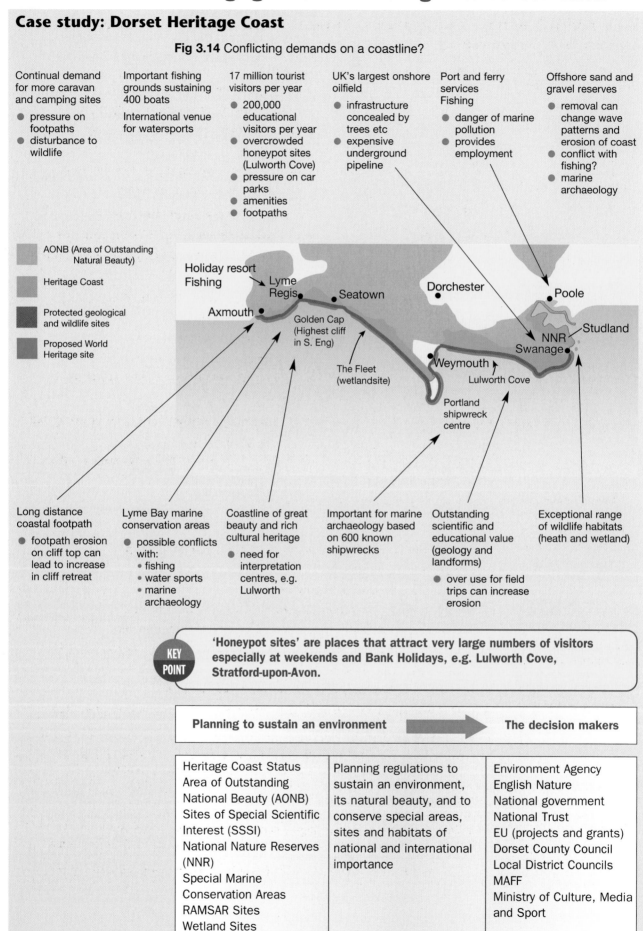

Continual demand for more caravan and camping sites
- pressure on footpaths
- disturbance to wildlife

Important fishing grounds sustaining 400 boats
International venue for watersports

17 million tourist visitors per year
- 200,000 educational visitors per year
- overcrowded honeypot sites (Lulworth Cove)
- pressure on car parks
- amenities
- footpaths

UK's largest onshore oilfield
- infrastructure concealed by trees etc
- expensive underground pipeline

Port and ferry services
Fishing
- danger of marine pollution
- provides employment

Offshore sand and gravel reserves
- removal can change wave patterns and erosion of coast
- conflict with fishing?
- marine archaeology

AONB (Area of Outstanding Natural Beauty)

Heritage Coast

Protected geological and wildlife sites

Proposed World Heritage site

Holiday resort
Fishing
Lyme Regis
Axmouth
Golden Cap (Highest cliff in S. Eng)
Seatown
The Fleet (wetlandsite)
Dorchester
Weymouth
Portland shipwreck centre
Lulworth Cove
Poole
Studland
NNR
Swanage

Long distance coastal footpath
- footpath erosion on cliff top can lead to increase in cliff retreat

Lyme Bay marine conservation areas
- possible conflicts with:
  - fishing
  - water sports
  - marine archaeology

Coastline of great beauty and rich cultural heritage
- need for interpretation centres, e.g. Lulworth

Important for marine archaeology based on 600 known shipwrecks

Outstanding scientific and educational value (geology and landforms)
- over use for field trips can increase erosion

Exceptional range of wildlife habitats (heath and wetland)

KEY POINT
'Honeypot sites' are places that attract very large numbers of visitors especially at weekends and Bank Holidays, e.g. Lulworth Cove, Stratford-upon-Avon.

Planning to sustain an environment		The decision makers
Heritage Coast Status Area of Outstanding National Beauty (AONB) Sites of Special Scientific Interest (SSSI) National Nature Reserves (NNR) Special Marine Conservation Areas RAMSAR Sites Wetland Sites	Planning regulations to sustain an environment, its natural beauty, and to conserve special areas, sites and habitats of national and international importance	Environment Agency English Nature National government National Trust EU (projects and grants) Dorset County Council Local District Councils MAFF Ministry of Culture, Media and Sport

Managing coasts for economic development

Case study: The Harris controversy, Outer Hebrides, north-west Scotland – a site for a superquarry

Fig 3.15 Possible superquarry site in the Outer Hebrides, Scotland

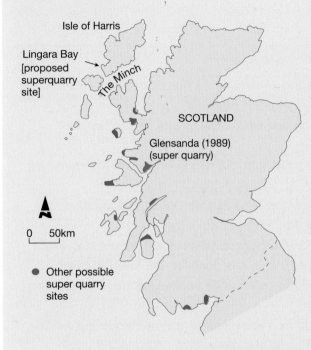

Isle of Harris

Lingara Bay [proposed superquarry site]

The Minch

SCOTLAND

Glensanda (1989) (super quarry)

0   50km

● Other possible super quarry sites

The need:

- sand, gravel and crushed rock are essential for most building projects, especially roads
- in Europe many millions of tonnes are required every year
- demand is exceeding supply from existing quarries; new sites are necessary.

The proposal for the Isle of Harris site:

- £50 million development of a superquarry
- $4km^2$ x 500m deep quarry, producing 10 million tonnes of rock per year.

Supporters – industry, Western Isles Council, majority of islanders
Supporters' argument:

- large reserve of suitable rock
- potential yield 600 million tonnes (60 years supply)
- deep-water harbour allowing direct loading onto bulk-carrying ships
- will provide more than 200 jobs, helping to reverse current depopulation of younger people
- very large income for local councils
- only a low population; little disturbance to local people.

Opposition – Scottish Natural Heritage, environmental groups
Opposition argument:

- will ruin the tranquil, wild environment
- will create an eyesore, visible from many kilometres away
- dust and noise pollution will affect local residents and wildlife
- increased danger of collision with oil tankers using the Minch
- quarried rock is a non-renewable resource
- the demand for crushed rock will lessen in the future
- if permission is granted, what will happen to all the other possible sites?

> **A decision from the Enquiry is still awaited. Look out for more information.**

> **Do you know of other case studies of coasts threatened by industrial, leisure, housing demands?**

PROGRESS CHECK

1. What name is given to the distance the wind has travelled over the sea?
2. Name the process of erosion when wave pressure breaks up rocks.
3. What is the most common landform produced by deposition?
4. Orford Ness is an example of which type of landform?
5. Is beach nourishment a 'hard' or 'soft' type of sea defence?
6. Which area of the UK has large areas at risk from coastal flooding?
7. What is the name given to a place that attracts very large numbers of visitors?

1. Fetch 2. Hydraulic action 3. Beach 4. Spit 5. 'Soft' 6. South east 7. Honeypot

Glacial landscapes

After studying this section, you should understand:

- that some landscapes in the UK are the result of the work of ice in the past
- how glaciers erode, transport and deposit material
- the formation of landforms associated with erosion and deposition by glaciers
- the aims of a National Park in the UK and the range of issues that must be considered
- the principal economic activities in an upland area, such as the Lake District, and their importance to the local community
- that the demands of the tourist and leisure industries are increasing and create physical and social problems in areas such as the Lake District
- that different groups make conflicting demands on the National Park, which have to be resolved
- that conflicts have to be resolved in a sustainable way with due regard to the needs of the local community and the aims of the National Park

Glacial landscapes

The spectacular landscapes found in upland areas of the UK such as North Wales, the Cairngorms and the Lake District, are mainly the result of the work of ice in the past. The 'ice age' consisted of many cold periods (**glacials**) separated by periods of warmer temperatures (**interglacials**).

- In very cold periods 30 000 years ago, **icesheets** extended over much of the UK, as far south as a line drawn from Bristol to London.
- When the last glacial period ended 10 000 years ago, only the valleys in upland areas were filled with **glaciers**, which flowed down from the highest mountains.

 KEY POINT Icesheets flow out of polar and upland areas to cover the whole landscape. Glaciers occupy valleys in upland areas, leaving the higher mountains exposed.

Movement of glaciers

- As ice accumulates and thickens it becomes heavier.
- The weight of the ice causes it to slide downhill.
- Movement creates friction with the rocks beneath the ice, which melts a thin layer of ice at the base of the glacier.
- The **meltwater** released acts as a lubricant and helps the glacier to move faster.

Glacial erosion

AQA A   AQA B   AQA C
EDEXCEL A
OCR C
WJEC

In cold glacial areas, freeze-thaw weathering is very active in breaking up rocks into loose angular fragments. There are two dominant processes of erosion:

- **Abrasion**
  - the ice scoops up loosened rocks as it moves forward
  - loose fragments fall onto the glacier from rocky slopes above
  - the moving ice uses this material to erode and scratch away the side and floor of the valley
- **Plucking**
  - ice freezes to the rocks beneath it
  - when the ice moves it pulls the rock apart along points of weakness

> Use these terms when explaining the landforms produced by erosion.

Landforms produced by glacial erosion

AQA A   AQA B   AQA C
EDEXCEL A
OCR C
WJEC

Corries (cirques or cwms) which are linked with:

- arêtes
- pyramidal peaks.

Glacial troughs (U-shaped valleys) which are linked with:

- hanging valleys
- truncated spurs
- ribbon lakes

Corries (Fig 4.1)

> You should be able to recognise these landforms on an OS map.

- deep, semi-circular hollows with very steep, precipitous sides or headwalls
- eroded from small hollows high up on the mountain sides, the hollows allow snow to accumulate, forming névé (compacted snow) and eventually dense glacier ice
- found in many upland areas such as North Wales e.g. Cadair Idris.

Arêtes are sharp, high ridges of rock where the headwalls of two corries are back-to-back.

Pyramidal peaks are sharp, angular mountain summits, formed by the erosion of several corries e.g. Cadair Idris.

Fig 4.1 Formation of a corrie

During a glacial period

Wind blown snow
Weight of snow forms glacial ice below
Ice - thinner: less erosion
Strong well-jointed rock
Névé
Headwall
Ice movement
Plucking
Rock lip
Abrasion
Plucking
Meltwater

Today

Freeze-thaw weathering
Headwall
Lake or tarn
Scree
Stream erosion

Glacial troughs or U-shaped valleys

- A valley glacier deepens, widens and straightens a former river valley.
- The V-shaped valley is eroded into a U-shape.
- The interlocking spurs are eroded and left as **truncated spurs** forming the steep valley sides.
- Tributary valleys which were not deepened as much as the main valley are left as **hanging valleys**, high up on the valley sides, marked today by waterfalls or steep gorges.
- Softer, less resistant rocks along the main valley floor were excavated more deeply and are now filled with water and called **ribbon lakes**.

Fig 4.2 Formation of a glacial trough

(a) Before glaciation

(b) During glaciation

(c) Today

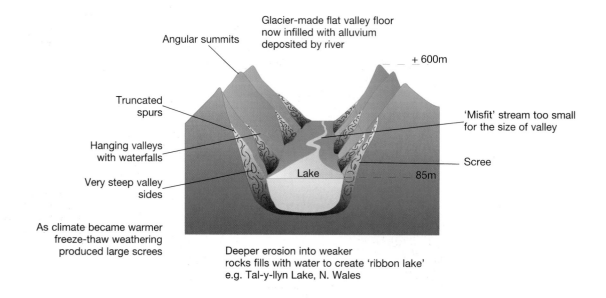

Case study of Cadair Idris, Wales

Fig 4.3a Block diagram of Cadair Idris

Fig 4.3b Outline map of the Cadair Idris area

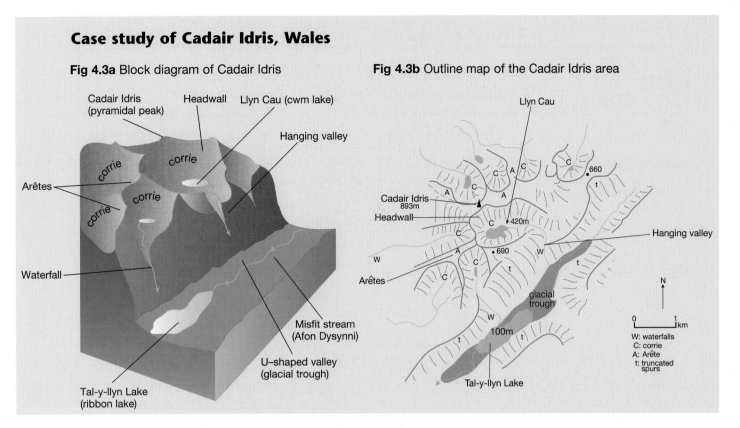

Transportation by ice

The large amount of rock carried by a glacier is known as its **load**. The rock carried by a glacier is called **moraine**. Moraine is carried within the glacier, on its surface, and at its base. Moraine consists of:

- very angular fragments of rock
- large boulders to fine clay.

Fig 4.4 Formation of lateral and medial moraines

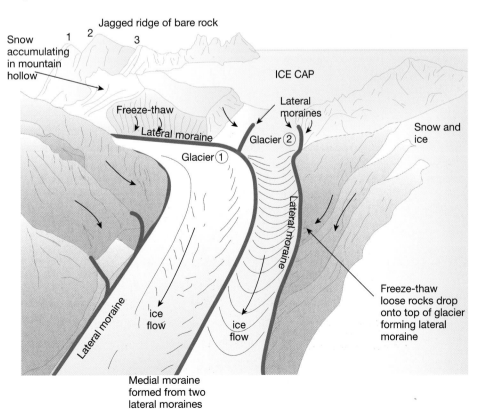

Landforms produced by glacial deposition

AQA A   AQA B   AQA C
EDEXCEL A
OCR A   OCR C
WJEC

The end of the last ice age was marked by:

- gradual increase of global temperatures and lower snowfall
- glaciers melted faster than the rate at which they moved forward
- glaciers became smaller, retreated and slowly disappeared from the UK.

> Glacial till is the more accurate and scientific term.

The moraine carried by the ice was left behind, deposited on the rocks and landscape below the ice. These deposits are known as **glacial till** or **boulder clay**. Glacial till is spread like a blanket over much of lowland UK, particularly in eastern areas such as Holderness, Lincolnshire and East Anglia.

Landforms formed of glacial till include:

Fig 4.5 Landforms of glacial deposition

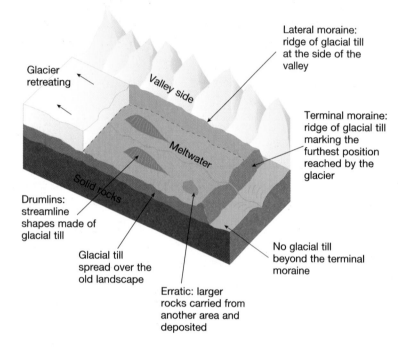

Lateral moraine: ridge of glacial till at the side of the valley

Glacier retreating

Valley side

Terminal moraine: ridge of glacial till marking the furthest position reached by the glacier

Meltwater

Solid rocks

Drumlins: streamline shapes made of glacial till

Glacial till spread over the old landscape

Erratic: larger rocks carried from another area and deposited

No glacial till beyond the terminal moraine

- **terminal moraines** – a low ridge of till, marking the furthest position reached by the ice front
- **lateral moraines** – low ridge of till, along the side of the valley floor, formed from rocks falling from the mountain sides above the glacier
- **drumlins** – streamlined shapes of till, often found in groups such as in the Eden Valley, Cumbria
- **erratics** – larger rocks transported by the ice and left in an area of very different rocks, e.g. rocks from Norway are found in East Anglia.

Human activities in an upland glaciated area

AQA A   AQA B   AQA C
EDEXCEL A   EDEXCEL B
OCR A   OCR C
WJEC

KEY POINT A National Park is a local government body administered mainly by local councillors and managed by an appointed National Park Officer and staff. (See also pages 176–8.)

Case study: the Lake District National Park

The Lake District National Park (see Fig. 4.8) is run by the Lake District National Park Authority (LDNPA) with the following aims:

- to conserve and enhance the natural beauty of the area
- to encourage and promote opportunities for people to enjoy leisure activities in the area and its natural beauty
- to manage and improve the quality of life of people living there (e.g. employment, housing, services).

Locate the Lake District in your atlas and identify some of these places.

Characteristics of the Lake District National Park:

- High mountain areas, such as Coniston Fell (800m), are separated from each other by deep, steep-sided, U-shaped valleys.
- Many valleys are partly filled by long, deep, ribbon lakes such as Coniston Water, Windermere and Ullswater.
- Much of the scenery and landforms are the result of the work of ice and glaciers in the past.
- Very high rainfall (Coniston has 2200mm/year compared with London which has 600mm/year) and impermeable rocks give rise to many tarns, mountain streams, waterfalls and lakes.
- The climate is characterised by cool summers (15°C compared with London, 22°C in July) and mild winters (5°C compared with London, 7°C in January).
- The most important land-use is pastoral farming, raising sheep and cattle.
- Large upland areas and steep valley sides are planted with forests for commercial use; 75 000 tonnes of timber are cut and sold each year.
- Communications are difficult; main roads mostly follow valleys.
- Tourism is very important and increasing (1998: 12 million visitors/year)

Fig 4.6 Sketch cross-section of the Coniston Valley

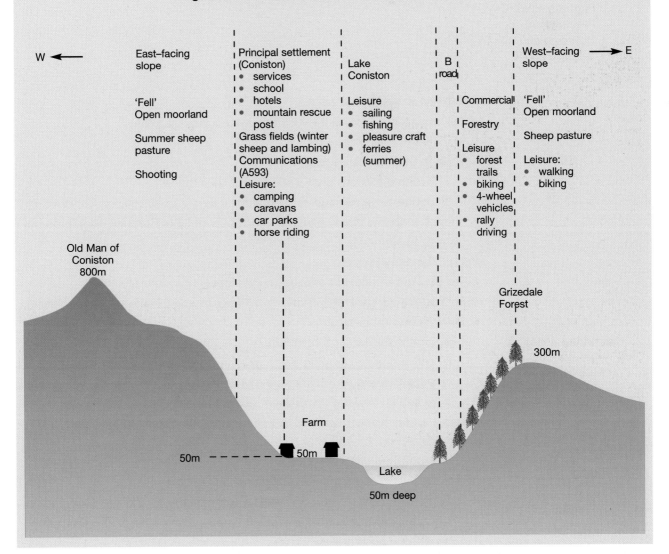

Case study: Hill farming in the Lake District

Hill farming produces lambs and calves. These are sold on to lowland farms in other areas of the country for fattening (meat) and breeding.

Fig 4.7 The farm system

Inputs	Processes	Outputs
Physical: • High rainfall • Low sunshine hours (cloudy) • High fells (upland areas) • Thin, poor soils on the fells • Steep valley sides • Limited, wet, flat land **Human:** • Family labour • Skilled knowledge **Economic:** • Low investment (poor profits, uncertain future) • High cost of supplementary feed • Distance from markets and difficult access – high fuel costs • Government policies: High subsidies (CAP) Price support • High cost of fertilisers • Veterinary fees	• Lowland grass management sowing mowing • Silage making (winter feed from grass) • Sheep rearing (lambing) • Cattle rearing (calving) • Fertilising • Bracken control on fells • Maintenance (buildings, stone walls)	• Lambs • Calves

Problems with farming
● Hill farmers have found it difficult to make a liveable profit.
● Many farms have been sold.
● Support from subsidies has led to:
    ● over-production, over-stocking and soil erosion
    ● quotas on sheep and cattle rearing
● There is a decline in the demand and supply of labour (young people do not want to be farmers).

Possible solutions
● Sell the land to housing or tourism developers (if in a suitable site).
● Seek additional part-time employment (e.g. forestry, National Park wardens).
● Diversify into other areas:
    ● farmhouse bed and breakfast
    ● conversion to camping barns (planning permission, investment)
    ● pony trekking (investment, expertise)
    ● quadbike trekking (investment, expertise)
● Seek support to farm the land in a more environmentally-friendly and sustainable way:
    ● Environmentally Sensitive Areas (ESAs) with subsidies to compensate for lower stocking levels and use of less fertiliser
    ● management to preserve wildlife, stone buildings and public access
    ● Countryside Stewardship – subsidies for conservation and preservation of stone buildings etc.
    ● Woodland Grant Scheme – grants to plant and manage woodlands of native species

> If you have visited a hill farm on a field trip, revise your notes and use as a case study.

Tourism in the Lake District

AQA
EDEXCEL
OCR
WJEC
CCEA

Tourism is a very important part of the local economy. Over 12 million people visit the Lake District each year (compared with a local population of 42 000). The number of visitors is increasing.

Visitors are attracted to the Lake District to:

- enjoy the scenery and landscape
- enjoy the peaceful environment and fresh air
- escape crowds!
- enjoy a variety of leisure pursuits, e.g. walking, water sports, rock climbing, mountain biking
- enjoy the attractive villages and shopping, e.g. craft shops, leisure wear shops

Fifty per cent of visitors come from:

- regions surrounding the Lake District, e.g. Newcastle, Manchester, Leeds
- areas made more easily accessible by the M6, e.g. Birmingham.

Ninety per cent of visitors travel by car – traffic has increased by over 30% in 20 years.

Visitors come for day visits or stay for weekends and longer holidays:

- about 30% stay in hotels and B&B accommodation
- about 40% rent cottages or bring caravans or tents.

In recent years some controversial proposals include:

- hotel extensions and leisure complexes on lake edges
- timeshare complexes
- marinas
- cable cars (proposal for Helvellyn)
- ski-lifts

Fig 4.8 Ten million people live within a three-hour drive of the Lake District

The impact of tourism

Local employment

- In some areas, 50% of the workforce is employed in the tourist industry, e.g. in Windermere: hotels, B&B, shops, cafes, leisure activities (boating).
- Many more jobs rely partly on tourism for business, e.g. builders, bakeries.
- Local tourism employment has helped to compensate for the decline of employment in agriculture and forestry.

Fig 4.9 Pie graphs comparing employment in the Lake District with the UK as a whole (1991)

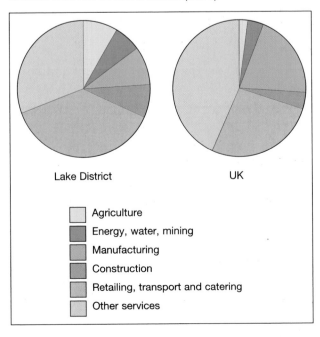

Study the pie graphs (Fig 4.9) and make comparisons under the headings:
•primary industry
●secondary industry
●service industry.

Challenges and problems:

● relatively low wages
● **seasonal employment** (up to 50% of jobs)
● **casual labour** often employed from outside Lake District (e.g. students)
● restricts development of skills and career
● many young people leave for employment in urban areas, e.g. Manchester.

Solutions:

● policy to recruit staff from local communities
● use local producers and services, e.g. builders
● extend tourist season – guided tours for retired people, discount weekends, theme weekends independent of weather
● encourage local craft industries to develop national markets, e.g. pottery
● encourage 'footloose' industries.

Local services

● Employment keeps people in the Lake District, which maintains local services such as schools, shops and leisure centres.
● Loss of 'local' shops to more profitable 'tourist' shops, e.g. gift shops, cafes, leisure wear (Grasmere and Ambleside).

Questions often ask for the benefits and problems faced by a local community as a result of tourism.

Housing

● Housing is in short supply. New housing is restricted to 'infilling' of existing settlements. There is a demand for **second homes**, often small cottages, and people are looking to invest in holiday letting. There is also a demand from outsiders for retirement homes.
● House prices have increased a great deal.
● Local people, particularly young people, cannot afford to buy so move out **(depopulation)** with a consequent threat to local services.

Solutions to housing problems:

● planning permission for new housing only granted if large proportion are only sold to local people
● **housing associations** build specifically for young couples.

Traffic and car parking

● The narrow, winding roads within the Lake District are heavily congested in summer, particularly at weekends.
● Car parking in towns, beauty spots and popular walking spots, e.g. Helvellyn, are at saturation point.
● Traffic and parking problems lead to inconvenience to locals, visitor frustration, illegal and dangerous parking.

Solutions to traffic and parking problems:

● restrict number of cars entering the National Park
● enforce more 'park and ride' schemes into popular areas where there is a single access, e.g. Haweswater
● build speed ramps in villages and restrict coaches, e.g. Grasmere.

Pollution

- There is pollution from cars and coaches particularly when the roads are congested.
- High visitor numbers leads to increased litter.
- There is an increase in sewage effluent in the summer.

Solutions to pollution problems:

- increase recycling in waste management
- educate visitors ('take litter home').

Physical environment

Fig 4.10 Stages in footpath erosion

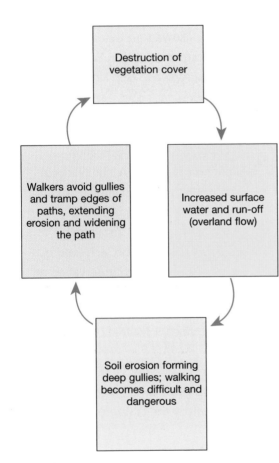

Footpath erosion

- Many access paths to popular mountain tops are so badly eroded that the scars spoil the scenery and the view from below – some scars are visible on satellite images!
- The amount of footpath erosion is influenced by:
  - human factors
    - number of walkers
    - proximity to car parks
    - winter use
  - physical factors
    - amount of rain
    - vegetation cover
    - angle of slope
    - soil compaction
- Repairs and maintenance are expensive in high, remote areas, e.g. £150/metre.

Solutions to problems of footpath erosion:

- obtain grants to carry out repairs to:
  - sympathetically restore the landscape, using local stone
  - provide a safe, hard-wearing path
- recognise sites liable to erosion in the future, and manage them carefully, e.g. diverting paths so they recover, fencing paths, improving drainage
- educate walkers.

There are also problems and conflicts in the rapidly-growing areas of mountain biking and 4-wheel-drive vehicles (Table 4.1).

Table 4.1 Issues of mountain biking and off-road leisure vehicles (4-wheel drive)

Mountain biking		Off-road leisure vehicles	
Problems	Solutions	Problems	Solutions
Rapidly growing activity – great demand	Establish more routes at low levels	Increasing ownership but limited knowledge of driving techniques	Establish a code of practice
Increases erosion of tracks and paths	Separate bikers from walkers	Many tracks made impassable	Classify routes based partly on difficulty
Conflict between bikers and walkers/farmers	Better signposting and education (cycle users code)	Extra noise pollution	More training

Use of lakes for leisure

- The increased use of lakes for water skiing, power boating, jet skiing has increased erosion of the lake edges.
- Conflicts have arisen between power boats, sailors, anglers, swimmers.
- There are problems of pollution and noise.

Solutions to problems associated with using the lakes for leisure:

- restrict use – water skiing now only allowed on Lake Windermere
- segregate uses, e.g. Lake Bassenthwaite (see Fig 4.11).

Sustainable tourism

Partnership and co-operation between the LDNPA, the National Trust, the hotel and leisure industry, and conservation groups are encouraging more sustainable tourism. This may be achieved by:

- encouraging the conservation of the landscape (restricting the number of cars?)
- raising funds to conserve and restore the landscape (entry fees to the Lake District?)
- raising awareness of visitors (more visitor centres, wardens?)
- showing that tourism and conservation must support each other if they are to be successful (too many people and cars destroy the features that attract the tourists)
- ensuring that tourism and conservation benefit the local community (employment, homes, services).

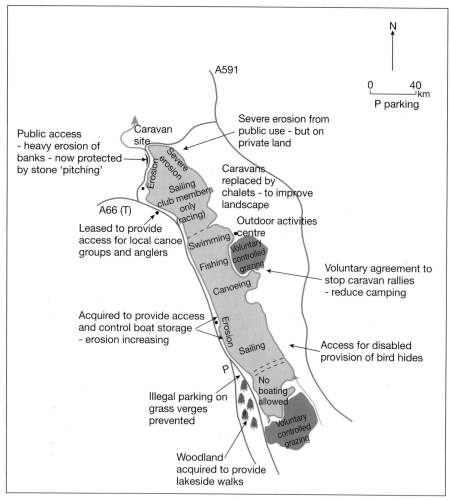

Fig 4.11 Management issues around Lake Bassenthwaite

Public access - heavy erosion of banks - now protected by stone 'pitching'

Severe erosion from public use - but on private land

Caravans replaced by chalets - to improve landscape

Outdoor activities centre

Voluntary agreement to stop caravan rallies - reduce camping

Access for disabled provision of bird hides

Voluntary controlled grazing

Leased to provide access for local canoe groups and anglers

Acquired to provide access and control boat storage - erosion increasing

Illegal parking on grass verges prevented

Woodland acquired to provide lakeside walks

Caravan site

Sailing club members only (racing)

Swimming

Fishing

Canoeing

Sailing

No boating allowed

Voluntary controlled grazing

A591

A66 (T)

N

0 — 40 km

P parking

1. Which weathering process is very active in glaciated areas?
2. What is the name given to sharp, high ridges of rock where the headwalls of two corries are back-to-back?
3. What are drumlins made of?
4. What is the name given to the ridge of morainic material marking the furthest extent of a glacier?
5. Name one of the outputs of a hill farm in the Lake District.
6. What industry offers the main employment for people living in the Lake District?

1. Freeze–thaw 2. Arête 3. Glacial till or boulder clay 4. Terminal moraine 5. Lambs or calves 6. Tourism

Weather and climate

LEARNING SUMMARY

After studying this section, you should understand:

- *the differences and relationships between weather and climate*
- *that solar radiation is the main source of energy in the atmosphere and that variations in solar energy over time (seasons) and space (latitude) are important in explaining weather and climatic patterns*
- *how atmospheric conditions can be measured and the processes explained*
- *how depressions and anticyclones affect the weather of the UK*
- *the factors affecting climate*
- *how to describe and compare climates in different parts of the world*
- *how weather and climate are affected by local conditions (microclimates)*
- *the formation and effects of tropical storms*
- *that atmospheric conditions may be changing as a result of human activities, which will have important social, economic and political consequences*

Weather

AQA A   AQA B
EDEXCEL A
OCR C
WJEC   Av.Hil
NICCEA

KEY POINT

Weather describes the atmospheric conditions at a particular place and time together with the changes taking place over the short term (hour-by-hour or day-by-day).

Recording the weather

Data on the weather is obtained from:
- daily readings of instruments at a weather station
- automatic instruments in remote locations (e.g. ocean buoys)
- ships and aeroplanes
- satellites and radar images.

Data is recorded at least once every 24 hours. The information is plotted on maps or synoptic charts, then analysed to obtain a weather forecast. **Meteorologists** are responsible for recording and analysing the weather.

A range of data is recorded

Temperature
- Air temperature is recorded in the shade (in a Stevenson's Screen).
- It is recorded using a **maximum/minimum thermometer** (highest and lowest temperatures in the last 24 hours).
- It is recorded in degrees **centigrade** (Celsius) or **Fahrenheit**.

Fig 5.1 Heat received by insolation

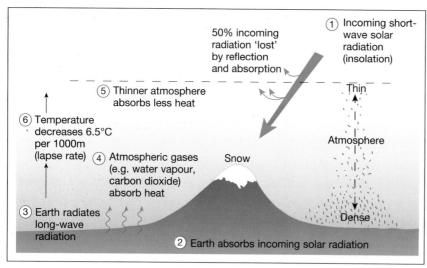

- Solar radiation or **insolation**, is the main source of heat energy to the atmosphere.
- The atmosphere is warmed mainly from **long wave radiation** emitted by the Earth.
- Temperature therefore decreases with height at an average rate of 6.5°C per 1000m (**lapse rate**).

Air pressure

- Air pressure is the weight of the atmosphere at the Earth's surface.
- It is recorded using an **aneroid barometer**.
- It is measured in **millibars (mb)**, reduced to sea level (average pressure at sea level is 1013mb).
- Air pressure varies with:
  - temperature: warmer, lighter air is forced to rise creating areas of low pressure; heavier, colder air descends creating areas of high pressure
  - height – as the atmosphere becomes thinner pressure decreases.

Fig 5.2 Causes of air pressure

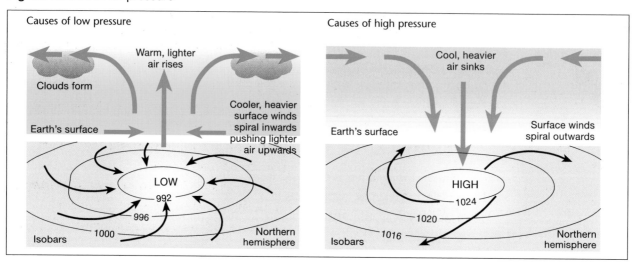

Fig 5.3 Giving wind directions

Wind

- Wind is the horizontal movement of air across the Earth's surface.
- It is recorded as:
  - direction using a **wind vane** (recorded as the direction from which the wind is blowing)
  - speed using an **anemometer**, measured in **knots**, **kilometres per hour** or on the **Beaufort Scale**.
- Wind is caused by differences in air pressure, the wind flows from high pressure (sinking) to low pressure (rising).

The greater the difference in pressure, the stronger the wind blows.

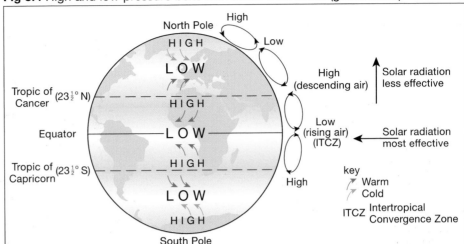

Fig 5.4 High and low pressure belts with associated winds (global scale)

> Recognising the global pressure and wind patterns is important when explaining the climate in different places.

Precipitation

- Precipitation is the deposition of moisture from the atmosphere as rain, drizzle, hail, snow, fog or dew.
- It is recorded using a **rain gauge**.
- It is measured in **millimetres (mm)**. (Snow is recorded as rainfall equivalent.)
- The atmosphere contains **water vapour**, which, if the air is cooled sufficiently, will form visible water droplets.

> Precipitation mainly occurs when air is forced to rise.

| Air cools and reaches saturation or **dewpoint** | → | Water vapour will **condense** to **liquid** | → | Visible water droplets form:
• **clouds**, if the air is forced to rise
• **fog** and **dew** if the air cools close to the ground |

Fig 5.5 Types of rainfall

A Relief rainfall

Types of rainfall:
- relief rainfall (orographic) – air is forced to rise over hills or mountains
- convectional rainfall – air, heated by the hot ground, is forced to rise
- frontal rainfall – when two air masses of different temperatures meet; the warm air mass is undercut and forced to rise by the heavier cold air.

B Convectional rainfall

C Frontal rainfall

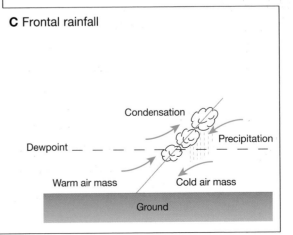

Fig 5.6 Types of clouds

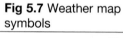

Cirrus	Cumulo-nimbus	Cumulus	Stratus
Very high wispy clouds	Very deep storm clouds	Individual fluffy clouds	Flat grey layers of cloud can form at any height

10 000m
d.p.

d.p.

d.p.

d.p.

d.p.

d.p. = dew point

Earth's surface

Clouds

- Clouds are visible masses of water droplets or ice crystals.
- They are recorded by observation; clouds are described by their shape and height.
- Cloud cover is measured in **oktas** (eighths of sky covered).

Sunshine

- Sunshine is recorded using a **sunshine recorder** and measured in **hours per day**.

Relative humidity (RH)

- Relative humidity is the amount of water vapour in the air.
- It is recorded by a **hygrometer** or **wet and dry thermometer**.
- It is measured as a percentage of the maximum water vapour that can be held by air of that temperature. It can be expressed in the following equation.

$$\frac{\text{water vapour present}}{\text{maximum water vapour that can be held by air of that temperature}} \times 100$$

- Warm air can hold more water vapour than cold air.
- If warm air is forced to rise and cool, RH will increase until the air becomes saturated, i.e. RH = 100% = **dewpoint**.

Fig 5.7 Weather map symbols

Station model

Cloud cover (4 oktas)

Temperature (4°C)

4

Wind speed and direction (15 knots N. West)

Precipitation (rain)

Cloud symbol	Cloud amount (oktas)	Symbol	Precipitation	Wind symbol	Wind speed (knots)
		=	Mist		
		≡	Fog		
○	0 No cloud	‚	Drizzle	○	Calm
◔	1 or less	;	Rain and drizzle		1-2
◑	2	•	Rain		
◕	3	✷	Rain and snow		3-7
◕	4	✳	Snow		
◕	5	✶	Rain shower		8-12
◕	6		Rain and snow shower		
●	7 or more	✱			13-17
●	8	✲	Snow shower	For each additional	
⊗	sky obscured	⬠	Hail shower	half-feather add 5	
		Ʀ	Thunderstorm	knots	

Weather maps or synoptic charts

KEY POINT Weather maps or synoptic charts summarise the weather data for a particular time.

- Data gathered from weather stations is represented as a group of symbols located at the site of the station.
- **Isobars** are drawn joining places of equal pressure.

Weather systems – depressions and anticyclones

Depressions

KEY POINT Depressions are areas of low pressure, formed when a warm air mass meets a cold air mass. Fronts form at the boundaries of the two air masses.

Fig 5.8A The formation of a depression

Stage 1: Warm and cold air meet – warm air rises

Stage 2: Low pressure develops, with warm and cold fronts

Stage 3: Fully developed frontal system formed as warm air rises

Stage 4: Front dies as cold air squeezes warm air upwards

Map

COLD POLAR AIR

B

A

Front

WARM MOIST TROPICAL AIR

LOW

COOL

A

B

WARM

Low pressure

COOL

A

B

Cold front moves forward faster than warm front

WARM

OCCLUSION

COOL

COOL

Cold front finally catches up with warm front and warm air is squeezed upwards

Cross-section

Warm air

COOLS

Front

Cool air

A

B

Warm air forced upwards

Warm front

Cold front

Warm air rises

COOL

COOL

A

B

WARM

COOL

COOL

A

B

FULL OCCLUSION

COOL

COOL

COOL

Depression forms over the North Atlantic

moves east

Fig 5.8B Cross-section of the formation of a depression

Cumulus

WARM AIR

Cirrus

Steeply rising air

Alto stratus

COOL AIR

Cumulo-nimbus

Nimbo stratus

COOL AIR

Clearing | Showers | Heavy rain | Cold front | Clear period | Warm front | Steady rain | Drizzle | High clouds increasing

Fig 5.8C Weather forecast for Norwich

'The weather for the next 24 hours will be very changeable. Clouds will rapidly increase with drizzle and then more persistent rain. It will become slightly warmer as the winds swing more southerly. Skies may clear briefly, before more heavy cloud and rain. There is a possibility of some thunder. Skies will slowly clear to give colder conditions.'

LOW

988

992

55°N

50°N

1000

1000

1004

Norwich

Movement of depression

- Depressions frequently form over the North Atlantic Ocean where warm, moist tropical air, moving north, meets cold polar air moving south.
- The depressions move east towards the UK, developing fronts (see Fig 5.8A).
- The cold air undercuts the warm air forming a cold front; the warm air is forced upwards.

Given a synoptic chart for the UK, be prepared to:
• describe the weather at a place given the weather symbols (Fig. 5.7)
• explain why it is raining close to a front that is shown (Fig 5.8b)
• provide a forecast for a place ahead of the depression (Fig. 5.8c)

- The warm air is pushed forward and forced to rise over the cold air ahead, forming a warm front.
- As warm air rises, clouds form and precipitation occurs along both fronts (see Fig 5.8B).
- Winds, in the Northern hemisphere, spiral inwards in an anti-clockwise direction. Winds will be strongest in deep depressions when the isobars are closer together.
- Meteorologists use their knowledge of depressions and fronts to make weather forecasts (see Fig 5.8C).
- Depressions are the main cause of the very changeable weather in the UK.

Anticyclones

> **KEY POINT** Anticyclones are areas of high pressure.

It is important to be able to explain the weather (see also Fig. 5.1).

- In anticyclones, the air is **sinking** and warming.
- Warmer air can hold more water vapour, so clouds are unlikely to form.
- Anticyclones can become stationary, giving settled weather for several days or weeks in the UK.

Fig 5.9 Anticyclones in summer and winter

Characteristic weather:

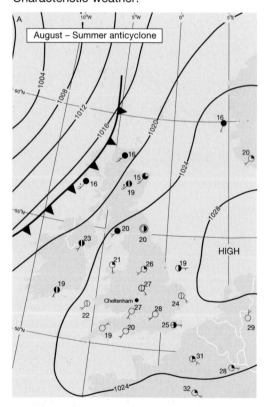

Average August temperature at Cheltenham: 17°C

Clear skies and light winds.
Temperatures fall overnight.

Average January temperature at Cheltenham: 4°C

Hot, fine weather:
Misty evenings and nights, which quickly clear in the morning.
Possible thunderstorms in late afternoon in inland areas of the country.
Above average temperatures may give a heatwave and drought conditions (UK 1990).

Cold, crisp weather:
Frosts and fogs at night are slow to clear in the morning and may last all day in inland areas of the country.
Possible snow showers along east coast.

Climate

AQA A   AQA B
EDEXCEL A
OCR C
WJEC   Av.Hil
NICCEA

KEY POINT Climate is the average weather conditions of a place or area, the result of weather data recorded over a long period of time (normally 30 years).

Factors affecting climate

Fig 5.10 The effect of latitude on climate: insolation

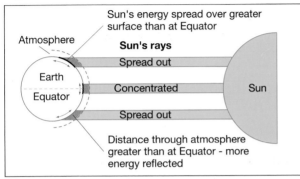

Latitude

With distance from the Equator:

- temperatures decrease as solar energy (insolation) is less concentrated towards the Poles
- seasonal differences in temperature are greater due to the effects of the tilt of the Earth's axis as it orbits the sun.

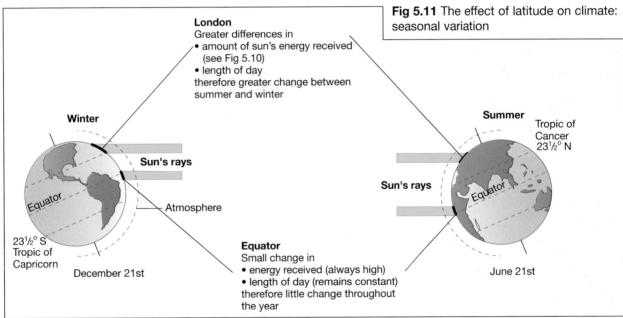

Fig 5.11 The effect of latitude on climate: seasonal variation

London
Greater differences in
• amount of sun's energy received (see Fig 5.10)
• length of day
therefore greater change between summer and winter

Winter
Sun's rays
Atmosphere
Equator
23½° S
Tropic of Capricorn
December 21st

Equator
Small change in
• energy received (always high)
• length of day (remains constant)
therefore little change throughout the year

Summer
Tropic of Cancer 23½° N
Sun's rays
Equator
June 21st

Altitude

Temperature falls, on average, 6.5°C for every 1000 metres in height (see Fig 5.1).

Distance from the sea

- The seas absorb and store heat to a greater depth than the land.
- The seas therefore warm up slowly in summer, storing heat which is slowly released through the winter. The land heats up more quickly but also cools more quickly.
- Coastal areas are kept cooler in summer and warmer in winter than land areas which are more distant from the sea.
- Islands and areas near the coast, such as the UK,

Fig 5.12 Altitude affects temperature

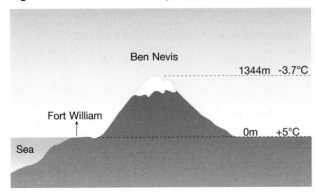

Ben Nevis
1344m -3.7°C

Fort William

0m +5°C

Sea

> Latitude, altitude, distance from the sea, prevailing winds and ocean currents are important factors affecting the climate. Refer to them when explaining the climate of any region.

have a **maritime** or **oceanic climate** with only a small range of temperature.

- Inland regions, such as central Germany, experience a more extreme **continental climate** with a large annual temperature range.
- The annual temperature range of two places at similar latitudes:
  London 13°C (see Fig 5.14); Berlin 21°C (see Fig 5.16B).

Prevailing winds

- Prevailing winds are the most frequent winds affecting an area.
- Winds blowing from the sea are generally warm and moist at all times.
- Winds blowing out from large land masses are generally hot and dry in summer and very cold and dry in winter.

Ocean currents

- Warm and cold currents move large distances across oceans.
- Currents moving from the Equator towards polar areas bring warm conditions to coastal areas further north.
- Currents moving south from polar areas carry cold water south causing sea fogs particularly in summer.

Fig 5.13 The Labrador Current and the NAD

Climatic regions

- At the global scale, the continents can be divided into climatic regions.
- These are extensive areas which have similar characteristics of temperature and precipitation.
- **Climate graphs** illustrate the main characteristics of each region.
- Maps show the extent and pattern of climatic regions. They illustrate the influence of:
  - latitude
  - ocean currents and prevailing winds
  - altitude.
- Within a climate region there will be many local variations of climate.

Fig 5.14 Climate graph for London (51$\frac{1}{2}$°N): cool, temperate maritime climate

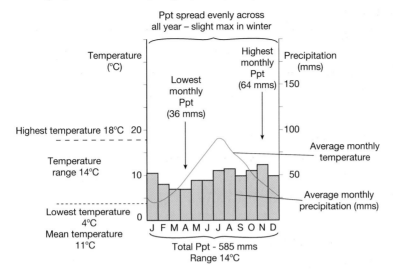

Fig 5.15 A climate map

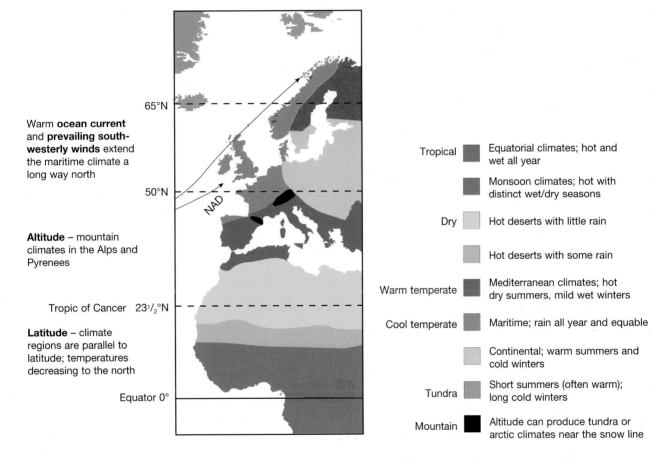

Characteristics of climates in Europe

Fig 5.16A-C Climate graphs for Shannon, Berlin and Rome

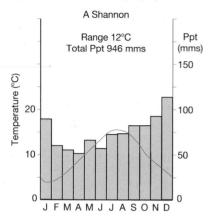

A Shannon

Range 12°C
Total Ppt 946 mms

A Shannon – cool, temperate maritime climate

- small temperature range, no extremes
- mild winters (5°C); cool summers (16°C)
- high precipitation; wet all year (no month below 50mm); slight winter maximum

Explanation:
- south westerly prevailing winds bring warm, moist, tropical maritime air
- closeness to the sea keeps summers cool but NAD ensures mild winters
- local uplands give relief rain all year
- depressions more frequent in winter – increase in frontal precipitation

B Berlin

Range 21°C
Total Ppt 618 mms

B Berlin – continental climate

- large temperature range (9°C greater than Shannon)
- short hot summers; very cold winters (Jan. -1°C)
- lower precipitation all year with summer maximum

Explanation:
- distant from influence of sea; land heats up and cools down quickly
- easterly winds in winter bring cold, dry, polar continental air – clear skies, hard frosts
- strong summer heating results in convectional storms and heavy rainfalls; winter snowfalls are common

C Rome

Range 17°C
Total Ppt 747 mms

C Rome – Mediterranean climate

- large temperature range but temperatures higher all year than Shannon and Berlin
- long, hot summers (max. 9°C higher than Shannon); mild winters (min. temp. Jan. 8°C)
- medium rainfall; summers very dry (July only 15mm, five times less than Shannon)

Explanation:
- lower latitude; sun's heat more intense, particularly in summer
- hot, dry, tropical continental air in summer – clear skies, maximum sun, little rain
- winter precipitation mostly frontal; high Alpine mountains to the north give protection from cold, polar, continental air in winter

Use precise data from the graphs to support your description of a climate.

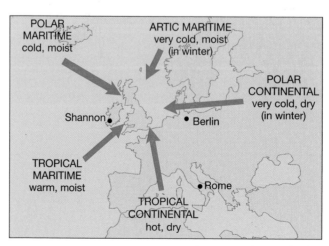

Weather and climate patterns in the UK

Fig 5.17A Seasonal temperature changes in the UK

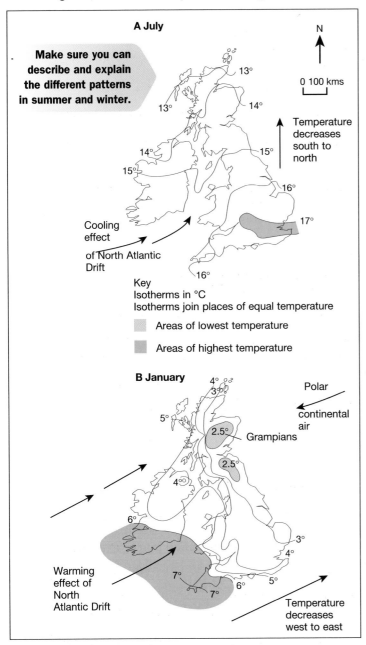

A July

Make sure you can describe and explain the different patterns in summer and winter.

N

0 100 kms

13°
13°   14°
14°   15°
15°
16°
17°
16°

Temperature decreases south to north

Cooling effect of North Atlantic Drift

Key
Isotherms in °C
Isotherms join places of equal temperature

Areas of lowest temperature
Areas of highest temperature

B January

4°
3°
5°
2.5°   Grampians
2.5°
4°
6°
7°   6°   5°
7°

Polar
continental air

Warming effect of North Atlantic Drift

Temperature decreases west to east

- Summer temperatures are warmer than winter temperatures (the south east is 13°C warmer).

  Explanation: changes in solar energy due to tilt of Earth's axis as it orbits the sun (see Fig 5.11).

- In summer, temperatures are highest in the south east (17°C) and coolest in northern Scotland (13°C).

  Explanation: solar energy is less concentrated at higher latitudes (see Fig 5.10). Northern areas more cloudy therefore less sunshine.

- In winter, temperatures in the south west (6°C) are 3.5°C warmer than in north east Scotland.

  Explanation: warming effect of the NAD (see Fig 5.13); cold polar air often reaches the north east.

- Precipitation is highest in western areas (+1600mm) and decreases further east (less than 600mm).

  Explanation: depressions approaching from the west bring large amounts of frontal precipitation (see Fig 5.8). Mountains in the west (e.g. Lake District) force south westerly prevailing winds to rise resulting in relief rainfall; drier eastern areas are in the rain shadow (see Fig 5.5A).

- In western areas, most precipitation falls in the winter; in eastern areas more rain falls in the summer and autumn.

  Explanation: there are more depressions in winter; summer heating in eastern areas results in convectional storms (see Fig 5.5B).

Fig 5.17B Precipitation patterns in the UK

Isohyets in mm of precipitation
Isohyets join places of equal precipitation

■ Areas over 1600mm

800
800
800   800
800   800
800   600
800
800

Area of lowest precipitation (less than 600mm)

South westerly prevailing winds

Precipitation decreases west to east

Fig 5.18

NW
Mild winters for the latitude (4°C); cool summers (12°C)
High precipitation all year; low sunshine hours
Most precipitation in winter; many gales
Large areas of high wet moorlands; acid soils
Hill sheep farming; summer tourism

NE
Cold winters (2°C); mild summers (12°C)
Low precipitation all year; often falling as snow in winter (Cairngorm Mountains have 100 days with snowfall per year)
Large-scale cattle rearing; some specialist arable farming along coast
Winter ski resorts in Cairngorms

SW
Mild winters (6°C); warm summers (16°C)
High sunshine hours in summer
High precipitation; wet winters, particularly in Welsh mountains
Dairying, cattle and hill sheep farming; specialist arable farming in Cornwall
Summer tourism; the south west (Devon and Cornwall) is a popular retirement venue

SE
Cold winters (less than 4°C), warm summers (16°C)
More extreme temperatures than SW
Lowest precipitation in UK; small summer maximum
Intensive arable farms in east; more cattle inland

Fig 5. 19 Climate graph for Uaupes (1°S): tropical equatorial climate

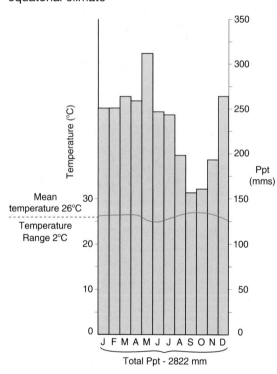

Total Ppt - 2822 mm

Contrasting climates

Uaupes, north west Brazil

- This climate occurs between 5°N and 5°S of the Equator.
- There are high temperatures all year, average 26°C. **Explanation**: close to the Equator, solar radiation is very high all year (see Fig 5.10); high sunshine hours.
- There is a very small temperature range (2°C) with no summer or winter seasons. **Explanation**: minimal effect of tilt of the Earth's axis (see Fig 5.11).
- There is a large day–night variation in temperatures (diurnal) (10°C). **Explanation**: clear skies at night – Earth cools rapidly (radiation) – early morning mists.
- There is very high precipitation (2822mm), over 150mm every month, and no wet and dry seasons. **Explanation**: high solar heating all day results in strong convection currents and clouds building up, giving heavy rain and thunderstorms every afternoon (see Fig 5.5B).

Table 5.1A Comparison of Uaupes with London – climate characteristics (refer to Figs 5.14 and 5.19)

	London	Uaupes	Uaupes compared with London
Mean temperature	11°C	26°C	More than double
Temperature range	14°C	2°C	Seven times smaller, no seasonal change
Highest temperature	18°C	27°C	9°C warmer all year than the summer in London
Lowest temperature	4°C	25°C	21°C warmer than January in London
Total precipitation	585mm	2822mm	Much higher, nearly five times as great; has more rain from January to March than London receives all year

Table 5.1B Comparison of Uaupes with London – vegetation and human activity

	London	**Uaupes**
Soils and vegetation	Deep fertile soil; deciduous woodland	Very deep but poor soils, particularly if trees are removed; tropical rainforest
Agriculture	Large commercial, intensive arable farms; capital intensive	Shifting cultivation, subsistence; minimum investment; commercial logging; cattle rearing after forest clearance
Population	Densely populated; very urbanised	Low density; semi-nomadic tribes

Local or microclimates

EDEXCEL A **EDEXCEL B**
OCR C

Land and sea breezes

Land and sea breezes frequently occur in the UK on hot summer days. During the day, the breeze is on-shore. Reverse flows of air take place at night when the land cools more than the sea (Fig 5.20).

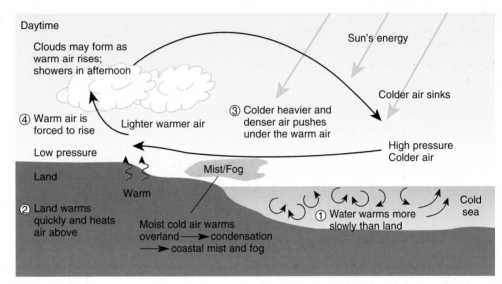

Fig 5.20 Land and sea breezes

Mountain and valley microclimates

Fig 5.21 Mountain and valley microclimates in the northern hemisphere

Urban microclimates

How is this different in a rural area?

- Temperatures in urban areas are higher than in the surrounding rural areas. Temperatures rise towards the centre of towns (urban heat island), e.g. central London is 6°C warmer than London Heathrow airport, 30 km away on the edge of the built-up area.

Explanation:

- concrete and tarmac absorb more solar energy than grass and trees
- dust and gases over urban areas reduce radiation, particularly at night
- central heating, air conditioning and traffic release heat into the air.
- There are high winds and turbulence in the streets particularly at night and in the winter. **Explanation**: wind is channelled down streets by high, vertical walls of buildings; turbulence increases with wind speed.

Weather hazards – tropical storms

AQA A   AQA B   AQA C
EDEXCEL A
OCR C
WJEC   Av. Hill

KEY POINT   Tropical storms are large low pressure systems which develop over oceans in the Tropics during the summer. They move north and south away from the Equator, along similar tracks. They are known in different parts of the world as cyclones, hurricanes, typhoons and willy-willies.

Tropical storms are hazardous particularly if they reach land because of:

- heavy rainfall (200mm per day) which can cause flooding and landslides damaging property and communications
- storm surges – low pressure causes the sea level to rise; large waves form in shallow coastal waters (+5m high). Low-lying coasts are very vulnerable, causing severe loss of life in highly populated areas
- high winds – often exceeding 200kph, cause severe damage to buildings, crops and electric wires.

Tropical storms form:

- over warm, tropical oceans with a water temperature of 27°C
- where high solar radiation causes rapid evaporation.

Stages in the development of a tropical storm

Fig 5.22 Cross-section through a tropical storm

Use the sequence shown, 1–8, to develop a logical explanation.

1. High solar heating; strong evaporation
2. Deep ocean; hot surface water (27°C)
3. Warm, very moist air rises quickly, cools and condenses
4. Very deep cumulo-nimbus clouds form; heavy rainfall
5. Low pressure develops; high winds as air is sucked into the depression
6. Rotation of the Earth encourages violent rotating winds around the central 'eye' (anticlockwise in northern hemisphere)
7. Central eye – descending air, clear skies, calm conditions
8. Whole system moves forward very fast in the prevailing winds (15-25kph)

Effects of tropical storms

MEDC case study: Hurricane Floyd

Fig 5.23 Hurricane Floyd menaces Georgia and the Carolinas

Alabama

South Carolina

Georgia

Charleston

Savannah

Florida

Interstate 26 massive traffic jams as Charleston evacuated

North and South Carolina 800,000 people evacuated

Hurricane Floyd reached a wind speed of 232 kph yesterday - a category 4 storm. Only two have ever exceeded this level. On "Labor Day", 1935 a storm killed 423 in Florida and Hurricane Camille killed 256 in Mississippi in 1969.

Kings Bay Naval Submarine Base 10 submarines sent to sea

Atlantic Ocean

Hurricane Floyd, 600 miles across, moving northwest at 23 kph

2 million Florida residents ordered to evacuate coastal areas

Gulf of Mexico

Orlando central cloud-free 'eye'

Disney World shut down

Hurricane began over the Atlantic as a circulating cloud system, powered by warm, moist air from the sea

Cape Canaveral 4 shuttles remain in hangers

Florida warplanes sent to Texas for safety

Miami

Miami Beach a ghost town

The Bahamas

Nassau

Bahamas pounded yesterday Roofs ripped off homes Flooding surged inland a quarter-mile at Nassau

> Locate the areas in Figs 5.23 and 5.24 on an atlas map. Are there any similarities in their location?

LEDC case study: The Bangladesh cyclone of 1991

Fig 5.24 The path of the cyclone

1. Path of cyclone funnelling into Bay of Bengal, winds gusting to 180kph
2. In shallower water approaching the coast, storm surge reaches 7m in height
3. Delta of the Ganges and Brahmaputra rivers is low-lying, most land is less than 1m above sea-level
4. Coastal defences consist of earth banks; many villages have been built on the drier banks
5. Dense rural population; few telephones; poor roads; some cyclone shelters
6. Storm surge swept away banks and villages; few people reached cyclone shelters. 12 000 died, 28 000 injured, crops and cattle washed away, 48 000 cattle lost, fishing boats destroyed
7. Roads and bridges swept away, rescue services poorly equipped
8. Food shortages and contaminated water; diseases such as cholera spread quickly

Prediction and protection

MEDCs:

- monitor and track storms by satellite and aircraft
- issue storm warnings at least 12 hours in advance
- prepare evacuation plans
- people prepare by:
  - building personal storm shelters
  - having adequate insurance cover
  - boarding-up windows
  - storing food and clean water supplies
- train emergency services

LEDCs:

- build cyclone shelters (£80 000 each), capable of safely housing a large number of families
- raise and strengthen earth banks along coast and rivers
- plant mangrove trees along coasts to build up silt and absorb storm surges
- educate people regarding risk
- obtain aid from MEDCs

> Keep up-to-date by keeping newspaper reports on current hazards and disasters.

> Be prepared to compare an MEDC with an LEDC in terms of underlying problems due to the difference in wealth. Compare the GNP of Bangladesh with that of the USA.

Climate change

AQA A AQA B AQA C
EDEXCEL A EDEXCEL B
OCR A OCR C
WJEC Av. Hill
NICCEA

Some scientists believe that the global climates are changing, largely because of human activity. Other scientists believe the changes being measured are only fluctuations in the atmospheric conditions, which have always occurred – ice sheets have advanced across the UK several times in the last 200 000 years (long before human activity had significant effects).

Evidence for change

> Articles in newspapers frequently discuss climate change when storms or flooding cause damage in the UK. Try to highlight the evidence for both sides of the argument.

- Rising global temperatures: average temperatures have increased by 0.6°C in the last 100 years – a further rise of 3°C is predicted by 2100.
- Extreme events are increasing: there are more intense tropical storms, and in the last few years the UK has had the driest, wettest and windiest periods of weather since records began.
- Changing ocean currents: currents in the South Pacific may be changing and causing droughts and forest fires in Indonesia and the loss of important fishing grounds off Peru and Chile.
- Glaciers and ice sheets are shrinking; permanently frozen soils and rock in Siberia and on high mountains in the Alps are melting.
- Various birds, fish and insects found in warmer regions in Africa are spreading north into Europe.

The principal environmental problems are: **global warming, ozone layer depletion, photochemical smog** and **acid deposition.** These problems require international solutions and cannot be dealt with by one country alone.

Global warming

> Frequently a subject of exam questions.

KEY POINT Global warming describes and explains the pattern of increasing global temperatures (see also pages 140–1.).

The greenhouse effect

Fig 5.25 The greenhouse effect

1 Solar energy reaches the Earth as short wave radiation

2 The Earth absorbs heat and radiates energy as long wave radiation

3 Greenhouse gases (e.g. water vapour, carbon dioxide, methane) absorb the long wave radation and insulate the Earth

4 Without insulation, the Earth would suffer extremes of temperature and would be uninhabitable

5 Greenhouse gases occur naturally but human activities have resulted in rising concentrations, more heat is retained and the atmosphere is becoming warmer

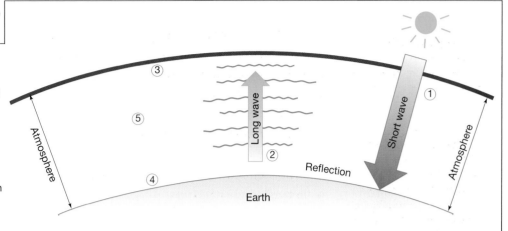

Causes

- **Fossil fuels** – the burning of fossil fuels, such as coal and oil, releases CO_2 into the atmosphere. The rate at which fossil fuels are burnt has increased rapidly in the last 100 years. Energy consumption in power stations, for domestic heating and in cars and lorries has increased by 50% in the last ten years.
- **Deforestation** – trees convert CO_2 into oxygen through photosynthesis. Most woodlands in temperate areas, such as the UK, have been chopped down. Now tropical rainforests are being felled. The burning of forests releases large amounts of CO_2 into the atmosphere.
- **Cattle farming** – cattle produce methane in their intestines; numbers of cattle have steadily increased as demand for food increases.
- **Waste tips** – decomposition releases large quantities of methane; commercial and domestic waste has increased enormously.

Fig 5.26 Areas at risk from sea level rise

Consequences

- A rise in sea-level:
  - melting of ice caps and glaciers will release large amounts of water into the oceans
  - higher water temperatures will cause water to expand
  - flooding of low-lying coastal areas such as Bangladesh, Japan and the UK Fenlands, and submerging of islands in the Caribbean and Indian Ocean will produce many millions of refugees.

- Global circulation will be disturbed:
  - tropical areas are becoming drier and deserts expanding which will lead to widespread crop failures in central Africa, malnutrition and starvation
  - water shortages will increase particularly in Africa, northern China and India
  - spread of malarial mosquito in Africa, USA and Europe
  - decreasing crop yields in South America, China, Africa and India

- more temperate conditions in Canada and Siberia will improve agriculture.

A few areas benefit but most will have problems.

- The melting of **permafrost** in Siberia will release large amounts of methane into the atmosphere and further increase global warming.
- Increased evaporation will increase amounts of water vapour in atmosphere and further increase global warming.
- Increased energy in the atmosphere will increase the likelihood of **extreme events** such as tropical storms.

Strategies to reduce global warming

- Reduce emissions of greenhouse gases:

Renewable sources of energy.

  - develop **alternative forms of energy** (solar, tidal, wind, hydroelectricity)
  - develop **alternative energy sources** for road transport (electric cars)
  - support LEDCs in using alternative forms of energy
  - agree international commitments to reduce emissions and limit pollution,

Keep up-to-date with 'new summits' which try to reduce pollution.

    e.g. **Earth Summits**
    - 1992 Rio de Janeiro (Agenda 21, 118 countries committed)
    - 1997 Kyoto (5% cut in emissions by all MEDCs)

Why are some countries unwilling to reduce their emissions of CO₂?

    - 2000 The Hague (USA is generally opposed to reductions, yet USA produces 5.5 million tonnes of CO_2 per year (1996), 5 times UK emissions and 230 times Bangladesh emissions)
- Reduce **deforestation** and develop sustainable policies such as **selective cutting** (rainforest) and **replanting** (new National Forest in the UK, in Leicestershire).
- Reduce population growth and reduce energy consumption.

Strategies to reduce impact of global warming

- Defend and protect low-lying coastal areas from flooding.

LEDCs will be more affected by drought and famine. How will these countries afford to pay for the research and new developments?

- Improve water supplies and use them more efficiently, particularly in areas that are becoming drier.
- Encourage research into developing crops to withstand drought and diseases.
- Develop more efficient ways of predicting and preparing for extreme weather events such as tropical storms.

Ozone layer depletion

 KEY POINT Ozone (O_3) forms a protective layer of gas around the Earth that absorbs and filters ultra-violet (UV) radiation from the sun.

Damage to the ozone layer

Do not give this as a reason for global warming.

- In 1986 scientists in Antarctica discovered a 'hole' in the ozone layer.
- Greater amounts of **UV radiation** are reaching the Earth.
- Measurements in 1998 showed that the hole had grown larger and amounts of **UV radiation** had increased in Australia and Argentina.
- A further hole was discovered over the Arctic, which is increasing in size and could affect Europe and North America.

 KEY POINT Ultraviolet radiation is an important component of sunlight which, in small doses, is good for people, producing vitamin D. However, excessive exposure can be harmful.

Causes

The increased production and use of **chlorofluorocarbons** (CFCs) which:

- are light gases which rise slowly into the atmosphere
- can have a long 'life' of over 400 years
- attack and destroy ozone molecules
- are responsible for slowly eroding a hole in the ozone layer.

> **KEY POINT**
> Chlorofluorocarbons (CFCs) are man-made chemicals used in foam packaging, aerosols and as a coolant in fridges and air-conditioning systems.

Fig 5.27 The ozone layer of the atmosphere

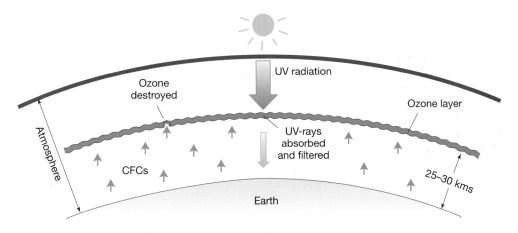

Consequences

In Australia and South America:

- skin cancer has increased; it is estimated that a 1% decrease in ozone causes a 5% increase in cancer
- eye cataracts have increased
- crops have been damaged by increased diseases
- marine food chains have been affected; micro-organisms that are important sources of food for fish are being destroyed.

> The Australian government and people treat this threat very seriously. 'Slap on the cream, slip on the hat.'

Solutions

- International Agreement in 1990 banned the use of CFCs by 2000:
  - most MEDCs have achieved this, some LEDCs require more time
  - new fridges no longer use CFCs, gases removed from old fridges, new 'ozone friendly' aerosols, biodegradable packaging has replaced foam
  - sufficient CFCs already in the atmosphere, however, to cause continued erosion of the ozone layer until at least 2020.
- Poster campaigns and the media have raised public awareness to use suntan creams and reduce exposure by always wearing sun-hats and long-sleeve shirts.
- Weather forecasts include information on UV intensities.

Photochemical smog

> **KEY POINT**
> Smog is a thick, ground level fog caused when water droplets become polluted by chemicals and gases.

> **Affects cities in MEDCs and LEDCs.**

Urban areas are experiencing an increase in the number of summer days with smog. Cities such as Calcutta, Mexico City, Athens and Los Angeles are badly affected.

Causes

Smog is more common:

● on clear, calm, hot summer days (higher UV radiation)
● where traffic is heavy and slow-moving (higher pollution levels)
● when cars are older with poorly maintained engines.

Fig 5.28 The formation of smog

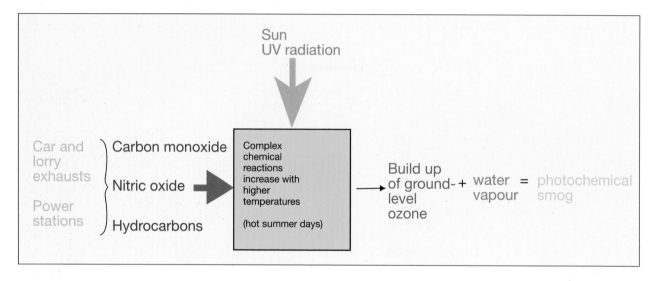

Consequences

● very unhealthy (increased hospital admissions and can be fatal to older people)
● causes irritation to the eyes and conjunctivitis
● causes breathing problems, increases asthma attacks
● detrimental to plants and animals

Solutions

● restrict cars in city centres, reduce lorry traffic
● improve and monitor engine efficiency
● improve electric-powered cars
● improve facilities for bicycles

Acid deposition

 KEY POINT   **Acid deposition is the fall-out from the atmosphere of acid compounds, either in rain as acid rain, or in very small particles as dry deposition.**

Acid rain was first recorded in Scandinavia during the 1950s. It is now recognised as affecting a large number of other countries including Canada, the USA, Germany, Scotland and Poland.

Fig 5.29

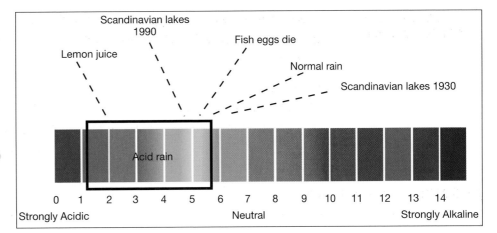

Acidity:
pH is a measure of acidity or alkalinity.

Causes

Fig 5.30 Causes of acid deposition

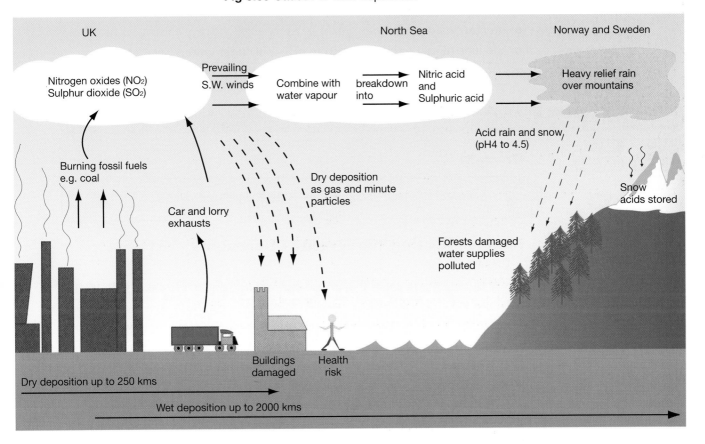

Consequences

- Acid rain releases nutrients in the soil which are carried away by percolating water (leaching); plants and trees suffer severe stress.
- Coniferous forests are dying; acid attacks the pine needles causing them to drop; greater risk from disease. 25% of the forests in southern Sweden, Germany and Poland have been severely damaged.
- Lakes and streams become so acid that they can no longer support any life at all, plants, fish or amphibians. (Meltwater from snow is very acid.)
- Lichens store chemicals which, when eaten by mice, rabbits and deer, contaminate the food chain (foxes, birds of prey).
- Acid soils restrict crop growth and reduce yields.

Affects both MEDCs and LEDCs.

- Health hazard – domestic water supply is contaminated; acid water attacks copper pipes making water toxic.
- Stonework on buildings attacked (chemical weathering – particularly limestone and marble) e.g. Houses of Parliament, Taj Mahal in India.
- Dry deposition can affect health: respiratory problems and headaches.

Solutions

- Increase height of power station chimneys – reduces local dry deposition but increases acid rain.
- Reduce SO_2 emissions from power stations by:
  - burning low sulphur coal – expensive to import
  - fitting filters to chimneys – expensive, could increase price of electricity
  - change to gas – increasing use of natural gas in recent years, and decrease in coal, have reduced emissions by over 40% since 1988 in the UK.
- Reduce nitric oxide emissions from cars by fitting catalytic converters.
- Use alternative energy sources – hydro, solar, wind.
- International Agreements.
- Add lime (alkali) to lakes to restore pH – only a temporary solution.
- Increase public awareness – reduce energy demand and use of cars.

Solutions are mostly the responsibility of the countries producing the pollution.

Drought and desertification

AQA A   AQA C
EDEXCEL A   EDEXCEL B
OCR A
WJEC   Av. Hill
NICCEA

Drought

KEY POINT — A drought is a long period of time without significant rainfall.

Factors affecting the severity of droughts

- Climatic factors
  - Past rainfall will affect the amounts of water in surface stores (rivers, lakes), and soil and groundwater (aquifer) stores (see Fig 2.2, page 18).
  - Rock type: permeable rocks (e.g. chalk) are able to store more water than impermeable rocks (e.g. clay – rapid run-off into rivers) (see Fig 1.13).
  - Temperature will affect the amount of water lost by evaporation.
- Human factors
  - The size of a country's population will affect the demand for water.
  - The economic wealth of a country will affect demand and its ability to cope with the problems (alternative water supplies).
  - Land-use: demands from agriculture (irrigation) and industry.

Causes of droughts in the UK

KEY POINT — An official drought in the UK is defined as a period of at least 15 consecutive days with 0.2mm of rain or less.

Typically in England it rains about one day in three; London receives on average 1.5mm per day throughout the year.

- Lower than average rainfall in spring and early summer.
- Summer anticyclones (high pressure systems) that become stationary over the UK for several weeks (see Fig 5.9) bringing clear, calm weather.
- Higher than average demand for water in hot weather.

Impact of droughts in MEDCs

- Water rationing by using street standpipes or tanker supplies (UK 1995).
- Reduced flows in rivers concentrate pollutants and reduce oxygen in water causing fish and aquatic plants to suffer or die.
- Reduction in yields of crops and grass, may lead to increased food prices.
- Damage to housing from shrinkage and cracking of clay soils.
- Increased fire risk in forests and heathland areas (e.g. New Forest, Hampshire). Can threaten property (e.g. California and Sydney 2000).
- Droughts are often associated with heatwaves (high pressure – see Fig 5.9) with risks of air pollution in urban areas (see Fig 5.28).

> Compare the impact of drought in MEDCs with that in LEDCs.

Managing the effects of drought

- Water conservation, banning hosepipes/washing cars, collecting rainwater
- Building new reservoirs
- Laying pipelines to transfer water from high rainfall areas to areas of high demand (e.g. Welsh water to Birmingham)

Impact of droughts in LEDCs

Case study: The Sahel

> **KEY POINT**
>
> The Sahel is belt of semi-arid land that stretches across Africa from Mauritania in the west to Somalia in the east. Since 1968 the area has suffered from many severe droughts.

Fig 5.31 Drought in the Sahel (see Fig. 5.4)

Many scientists blame climate change for disturbing the global circulation of winds and ocean currents and causing the rains to fail in the Sahel. Other scientists consider the droughts to be only short-term fluctuations of climate.

Desertification is also spreading north of the Sahara, in countries such as Algeria, Spain, Italy and Turkey.

Causes of drought

- Failure of the seasonal summer rains.
- With less rainfall, trees and grassland cannot survive and becomes scrub vegetation.
- Less vegetation gives less transpiration, so a smaller amount of water vapour is available and therefore there is less likelihood of rain.

Desertification

 KEY POINT   Desertification is the process by which land becomes desert.

Drought and human activities have led to the Sahel becoming more like a desert, with:

- thin, dry, sandy soils,
- bare rock where the soils have been blown away (soil erosion)
- sparse vegetation.

The processes of desertification

Fig 5.32 Desertification in the Sahel

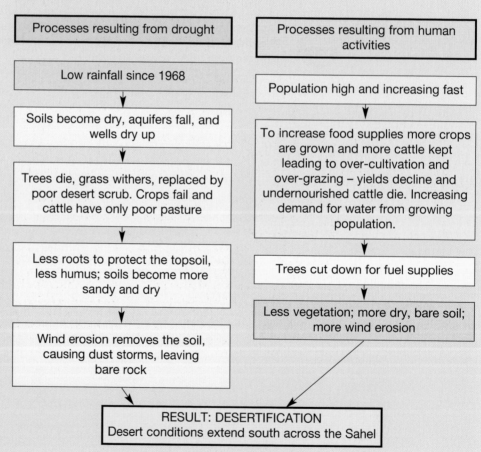

Effects of drought and desertification

- Food and water shortages lead to malnutrition, famine, disease and high death rates.
- Large numbers of people become dependent on food aid programmes.
- There is migration of people
  - from rural to urban areas causing over-population in towns
  - to refugee camps.

Solutions

- In the short term, provide food aid and water supplies to prevent suffering.
- Improve water supplies by building large reservoirs and drilling deeper wells
- Conserve water in local, small-scale schemes (e.g. building low stone walls across a slope – Burkina Faso).
- Improve farming methods by encouraging schemes using sustainable appropriate technologies (e.g. small-scale, locally-made tools, not tractors).
- Provide drought-resistant seed such as millet (northern Nigeria).
- Start tree-planting schemes to reduce soil erosion (Mauritania).
- International action to reduce the causes of global warming.

Population growth continues to outstrip food supplies.

Civil wars as in Eritrea, Ethiopia and Sudan prevent aid reaching stricken areas, and cause mass migrations to refugee camps.

PROGRESS CHECK

1. What is the difference between weather and climate?
2. Which instrument is used to measure air pressure?
3. What is the name given to a line that joins places of equal rainfall?
4. Is air rising or sinking in an anticyclone?
5. Why is the south west of the UK relatively warm in winter?
6. Which of the following areas of the UK receives the highest average precipitation per year:
   - the London area
   - the north-east
   - the Lake District?
7. What is the name given to the type of climate experienced in Berlin?
8. Why is the layer of ozone found in the upper atmosphere beneficial to life on Earth?
9. Which of the following are causing global warming:
   - the hole in the ozone layer
   - the increase in greenhouse gases
   - the rise in sea level?
10. What is desertification?

1. Weather describes atmospheric conditions at a certain time and place; climate is the average weather of a place over many years. 2. Barometer 3. Isohyet 4. Sinking 5. Influence of the warm ocean current called the North Atlantic Drift 6. The Lake District 7. Continental 8. Protects us from exposure to too much ultra-violet radiation 9. The increase in greenhouse gases 10. The process caused by natural and human factors that turns land into desert

After studying this section, you should understand:

LEARNING SUMMARY

- *that ecosystems describe the relationships between climate, soils and living organisms*
- *that ecosystems exist at a variety of scales*
- *how energy from the sun is circulated through the system by plants and animals*
- *how nutrients are derived from weathered rock and are circulated through the system by plants and animals*
- *that ecosystems develop over time by the process of succession*
- *how most ecosystems have been destroyed or altered by human activity*
- *that many ecosystems need to managed in a sustainable way*

The natural environment

AQA A   AQA B
EDEXCEL A   EDEXCEL B
OCR A   OCR B
WJEC   Av. Hill
NICCEA

The **natural environment** consists of the **rocks** and **soils**, **vegetation** and **animals**, and the **air** and **water** of the Earth; it is made up of different **ecosystems**.

> **KEY POINT**
>
> An ecosystem describes the links between the living community of animals and plants and their habitat.

- The living community is the **biomass** (animals, plants, fungi, bacteria).
- They are linked to each other in a **food chain**.
- The **habitat** includes the non-living components such as rocks, soil, water and climate.
- Ecosystems exist at a variety of scales, from a rock pool on the seashore to **biomes**, which are large global systems, such as the tropical rainforest.
- Human activity has destroyed or altered parts of most ecosystems.

Energy cycle

> *Understanding the basic ideas of the energy cycle, nutrient cycle and succession will help you to explain why biomes are so different.*

- The sun is the source of all energy in an ecosystem.
- Light energy from the sun is absorbed by green plants in a process called **photosynthesis**.
- Energy flows through the system as a food chain. Plants are eaten by animals, and animals eat each other.
- The food chain consists of a series of stages known as trophic levels.
- Energy is stored in the system: in plant tissue, muscle and fat.
- Energy is lost from the system by transpiration, respiration (by plants and animals), consumption (when a worm is eaten by a bird) and decomposition (bacteria and fungi break down dead plants and animals).
- Human activity can interfere with the food chain, e.g. deforestation, keeping large flocks of sheep, hunting large carnivores, introducing new species.

Fig 6.1 An ecosystem

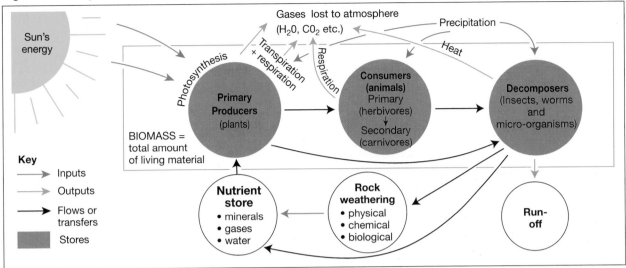

Fig 6.2 Trophic levels of an ecosystem

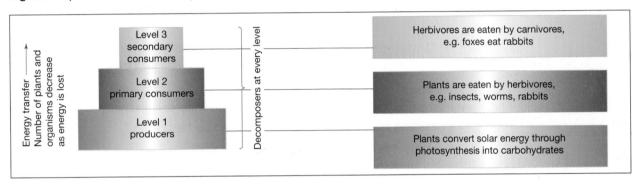

Nutrient cycle

- Weathered rock releases nutrients such as calcium, carbon and nitrogen into the soil.
- Plants and animals circulate nutrients through the system.
- The cycle occurs quickly in hot, wet ecosystems, e.g. tropical rainforest, and more slowly in cold systems, e.g. coniferous forest.
- Nutrients are lost by percolation and run-off into rivers.
- Human activity can affect the cycle by deforestation – loss of plant material and increased run-off.

Fig 6.3 The nutrient cycle

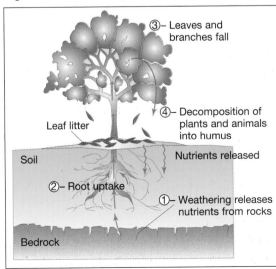

Succession

Ecosystems develop over time by the process of succession.

- The first plants to colonise an area are called pioneer species.
- Pioneer plants are adapted to harsh conditions. They are hardy and require only a low nutrient supply.
- Rock weathering and the decomposition of plants increase the supply of nutrients, allowing new plants to germinate, e.g. grasses. Insect and animal life increases.
- Soils become deeper, more nutrients – larger and more varied species, e.g. shrubs.

Fig 6.4 Typical succession to climax vegetation

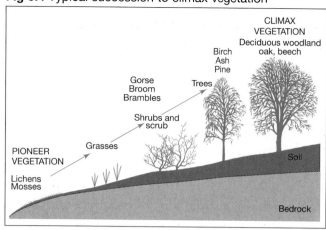

- Given time, a dominant species, such as oak or beech trees, take over and smaller plants live beneath the dominant trees. The succession is completed.
- Plants, insects, birds and animals exist in complex food chains and webs.
- There are four types of succession:
  - on bare rock
  - in freshwater
  - on sand
  - in salt water.

A small-scale ecosystem – sand dunes

- Dunes form along coasts where there is a large supply of sand from the beach and strong prevailing winds blowing in from the sea. Dunes are common on sand spits (see Fig 3.9).
- Sand is blown inland forming low embryo dunes which are colonised by marram grass. Further sand is trapped by the grass and the dunes grow higher.
- Marram grass is **adapted** to the dry and windy environment by:
  - growing fast so as not to be smothered by fresh sand
  - having long roots to reach the water table
  - having strong leaves that can stand exposure to high winds and prevent excessive moisture loss.

The plants and animals interact with the wind and sand, and the succession develops.

- Marram grass helps to stabilise the dunes – decomposition provides humus for soil development.
- Older dunes are more stable and have a wider variety of plants covering the ground. Shrubby plants develop (e.g. sea buckthorn, hawthorn, gorse) which are colonised by insects, birds, rabbits, birds of prey and foxes. Pine trees and eventually the dominant species of oak and ash complete the succession.

Fig 6.5 Succession to climax vegetation on sand dunes

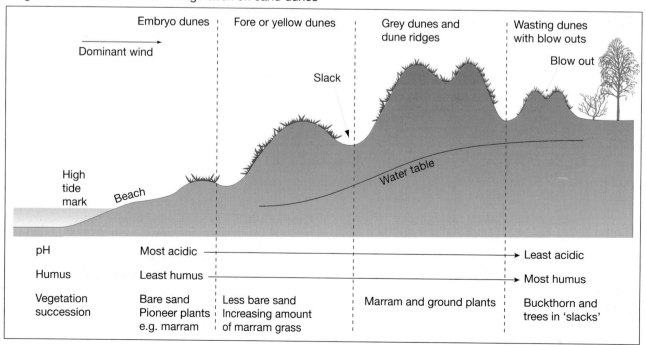

Dune management
- Dunes are important coastal defences and need to be conserved.
- Dunes are fragile systems and can quickly be damaged by:
  - trampling, sliding down dune faces, horse riding and fires
  - over-population by rabbits
  - military use.
- Damage to dunes leads to:
  - vegetation being destroyed
  - bare sand exposed
  - wind removes sand, leaving 'blow-outs', exposing root systems which kills marram grass
  - greater exposure of bare sand – more wind erosion.
- Management techniques include:
  - wooden boardwalks and fencing along popular routes through dunes
  - information/interpretative boards to raise public awareness
  - wardens and guides
  - car parking charges to reduce demand
  - culling and controlling rabbit populations.

> **Sand dunes at Gibraltar Point, Skegness (Fig 3.12) and Studland (Fig 3.14) are carefully managed to allow public access and sustain the dune habitat.**

Biomes

AQA A   AQA B
EDEXCEL A   EDEXCEL B
OCR A   OCR C
WJEC   Av. Hill
NICCEA

> **KEY POINT**   Biomes are large ecosystems at the global scale where the climate, vegetation and soils are broadly the same.

- Human activity has destroyed or changed parts of all biomes.
- There are eight major biomes:
  - Tundra
  - Mediterranean
  - Coniferous forest (taiga)
  - Desert
  - Temperate deciduous woodland
  - Savanna grassland
  - Temperate grasslands
  - Tropical rainforest

Fig 6.6 Four of the major biomes

Key
A = Archangel
S = Shannon
K = Kano
U = Uaupes

23.5° Tropic of Cancer
Equator
23.5° Tropic of Capricorn

Coniferous forests
Temperate deciduous woodland
Savanna grasslands
Tropical rainforests

Fig 6.7A Climate graph for Archangel

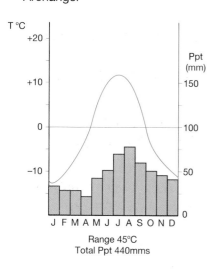

Range 45°C
Total Ppt 440mms

Coniferous forest

Location (see Fig 6.6):

- found only in the northern hemisphere in a wide band between 50°N and the Arctic Circle; found in USA, Canada, northern Europe and Siberia

Climate:

- long, cold winters (average temperature -15°C) with extreme temperatures (-30°C) and frosts. Short summers with long days – short **growing season** (three months)
- low precipitation (500mm) – cold air unable to hold much water vapour. Summer maximum, winter snowfall melts only in spring

Vegetation and animals:

- spruce, pine and fir: large areas of single species, few other plants owing to thick layer of pine needles and lack of light all year from tree cover
- **evergreen**, keeping leaves (needles) all year
- small needle-shaped leaves, waxy, to reduce transpiration
- conical tree shape
- seeds protected in cones
- voles, hares, reindeer, wolves, bears – few birds except in summer

Soils:

- **podsols**: shallow acidic soils; nutrients and minerals washed downwards (**leaching**) by percolating water in summer
- pine needles slow to decompose (low temperatures) – thin humus layer
- clear layers (**horizons**), few worms

> Note how the vegetation is adapted to the cold climate.

Fig 6.7B A coniferous tree

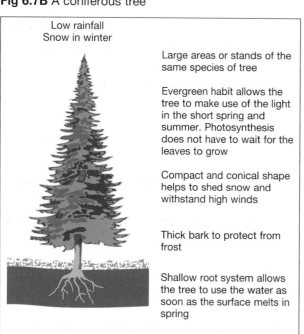

Low rainfall
Snow in winter

Large areas or stands of the same species of tree

Evergreen habit allows the tree to make use of the light in the short spring and summer. Photosynthesis does not have to wait for the leaves to grow

Compact and conical shape helps to shed snow and withstand high winds

Thick bark to protect from frost

Shallow root system allows the tree to use the water as soon as the surface melts in spring

Fig 6.7C Podsol soil profile

Thick litter layer breaks down slowly

More humus
pH 3.5 to 4.5
Sandy, well-drained soil
Ash grey in colour
Horizons very distinct
Strongly differentiated horizons – few soil animals
Leaching of iron minerals
Red colour
Nutrient accumulation

Hardpan can form (hard layer where clay is cemented together by iron minerals)

Acid parent rock

Human use:

- very productive because of large **biomass**; most forests commercially managed
- tall, straight trunks provide excellent softwood planks for building industry
- **softwood** provides wood pulp for paper industry

Fig 6.8A Climate graph for Shannon

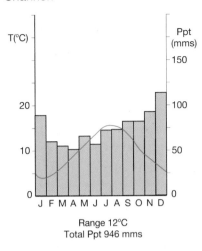

Range 12°C
Total Ppt 946 mms

Temperate deciduous woodland

Location (see Fig 6.6):

- mostly on the east coast of continents, between 40°N to 60°N, but also includes north-west Europe
- includes eastern USA, much of China, Japan and New Zealand

Climate:

- cool summers – growing season is seven months; mild winters
- precipitation over 1000mm/year, falling throughout the year

Vegetation and animals:

- **deciduous** trees (oak, beech, ash): shed leaves in winter (reduces transpiration when colder temperatures reduces effectiveness of **photosynthesis**)
- relatively few species compared with tropical rainforest
- oak represents **climax vegetation**
- there are some layers beneath the upper canopy as light can penetrate, particularly in spring (shrub layer, field layer)
- thick leaf litter, but many decomposers cause rapid recycling of nutrients
- deer, foxes, squirrels, badgers, many bird species

Fig 6.8B Deciduous woodland

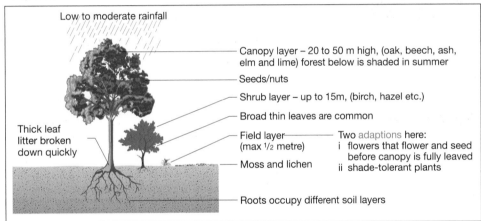

Low to moderate rainfall

Canopy layer – 20 to 50 m high, (oak, beech, ash, elm and lime) forest below is shaded in summer

Seeds/nuts

Shrub layer – up to 15m, (birch, hazel etc.)

Broad thin leaves are common

Field layer (max ½ metre)

Moss and lichen

Thick leaf litter broken down quickly

Two adaptions here:
i   flowers that flower and seed before canopy is fully leaved
ii shade-tolerant plants

Roots occupy different soil layers

Fig 6.8C Brown earth soil profile

Litter layer
Humus incorporated rapidly
Light grey and very porous
Worms and other soil animals mix soil and minerals in soil. There are poorly defined horizons in this soil
Ⓐ
Well drained
Nutrients well distributed
pH 5 to 5.5
Ⓑ
Grading through red to brown
Parent material
Ⓒ
30-50 cms

Soils:

- **brown earths** – deeper and more fertile than podsols
- leaf litter decomposes quickly and is incorporated into the soil by many earthworms
- fairly uniform, well-drained soil, deeply penetrated by root systems of trees

Human activity:

- severe deforestation, the result of clearance and use over the past 1000 years
- significant uses today include game management, timber production and recreation – multi-purpose seen as best way of conservation
- schemes to re-establish forest with multi-purpose uses:

- **Community Forest Organisation** – since 1991, 3500 hectares of native trees have been planted in twelve community forests in the UK, mostly on brownfield sites on the edges of cities, e.g. Great North forest (Newcastle), Mersey forest (Liverpool - Manchester)
- **National Forest** – designated in 1990 with aims to:
  - plant a third of the Midlands (Staffordshire, Derbyshire, Leicestershire) with the new woodland
  - use **brownfield sites** (formerly mining) for commercial forestry, leisure pursuits (e.g. cycling, fishing, walking) and wildlife reserves.

Fig 6.9A Climate graph for Kano

T (°C)
Range 12°C
Total Ppt 840 mms

Savanna grasslands

Location (see Fig 6.6):
- found between 25°N and 25°S of the Equator, mostly covers the central parts of continents
- includes central Africa (northern Nigeria, Kenya, Zambia,), north Australia and central South America (Venezuela, Brazil)

Climate:
- temperatures are high throughout the year (average 25°C) with a short cooler season
- **wet and dry seasons** – summer wet season when sun gives maximum heat (see Fig 5.11) – 80% of rainfall occurs in four months as heavy convectional storms
- little rain in dry season with dry winds blowing from the continent (see Fig 5.31)
- wet season may be becoming unreliable due to climatic change leading to **desertification** (see Chapter 5, page 75)

 KEY POINT Savanna is transitional between tropical rainforest (heavy rain all year) and desert (low unreliable rainfall)

Vegetation and animals:
- mixture of trees and grasses, with grasses dominating towards the desert
- deciduous trees, shedding leaves in dry season to limit transpiration and loss of water (e.g. baobab tree)
- trees have adapted to drought by having:
  - small, waxy or thorn-like leaves
  - large trunks capable of storing water
  - long roots, extending down to water table
  - thick, fire-resistant bark (fires are common as a result of lightning during convectional storms)
- grasses grow in tufts, with thick strong blades
- grasses grow quickly with the first rains, up to 5m in height, and die back in dry season; seeds remain dormant until rain falls again
- supports many species of **herbivore** such as zebra and antelope (Fig 6.10)
- supports **carnivores** such as lion and hyena which feed on herbivores

Fig 6.9B Savanna grassland section

Fig 6.9C Savanna soil profile

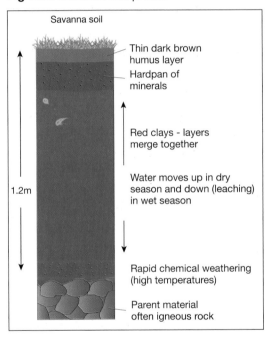

Soil:

- soils are red-coloured clays, rich in iron and aluminium oxides
- dead leaves and grass rapidly decay in high temperatures but give little humus or nutrients (low fertility)
- hard layer ('pan') develops just below surface, restricting penetration of water and roots – makes ploughing difficult and, if vegetation is removed, the soils are easily eroded by the wind.

Human activity:

- most suitable for nomadic pastoralism – extensive, low density cattle and sheep grazing, following the wet season (fresh grasses), north and south
- crops can be grown , particularly where land is irrigated – maize, millet and tobacco
- increased population and decline in nomadism (settling around water holes) lead to overgrazing – removal of vegetation and increased wind erosion and **desertification** (see Chapter 5, page 75)

Fig 6.10 Simple food web in savanna grassland

Fig 6.11A Climate graph for Uaupes, Brazil

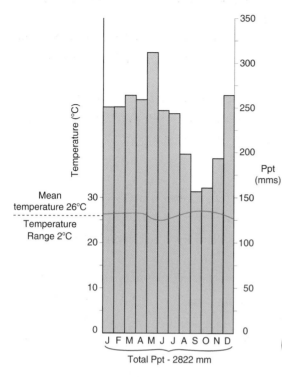

Mean temperature 26°C

Temperature Range 2°C

Total Ppt - 2822 mm

Tropical rainforests

Location (see Fig 6.6):
- found between 5°N and 5°S of the Equator
- areas include Amazon Basin (Brazil), West Africa (southern Nigeria, Congo), South East Asia (Indonesia) and north east Australia (state of Queensland)

Climate (see Chapter 5, page 64):
- high temperatures all year (average 26°C); small annual temperature range (2°C); 12 month continuous growing season
- annual rainfall very high (over 2000mm), falling throughout year (some areas slightly away from Equator have a short 'drier' season)
- high humidity and rising air gives daily convectional storms (in the afternoon)

KEY POINT   **The biome with the most diverse vegetation and animal life: Amazon rainforest has up to 300 species of tree per square km – mostly hardwoods, e.g. ebony, teak, mahogany**

Vegetation and animals:

- **deciduous** trees, growing and shedding leaves throughout year, gives evergreen appearance; flowers, fruits and seeds available all year from different species of tree – huge continuous supply of leaves to forest floor
- distinct layers of vegetation with the tallest trees up to 50m tall (**emergents**)
- trees have shallow roots but develop **buttress roots** above ground to support their great height
- leaves are thick and waxy to reduce transpiration if exposed to the sun; they have '**drip tips**' in order to shed the heavy rainfall efficiently

Fig 6.11B Cross-section through a tropical rainforest

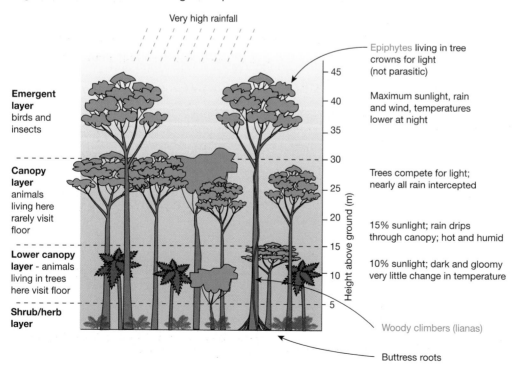

Fig 6.11C Tropical soil profile

- Plentiful leaf debris
- Rapid breakdown and recycling of forest debris
- Small amount of humus
- Red and brown colour
- Leaching of iron minerals

20 m

Increasing acidity

- Clay

Red
↓
Yellow
↓
White

- Rapid weathering of parent rock (hot and wet)

- **epiphytes** grow on trunks and branches, but do not feed on the trees' nutrients
- at ground level the forest is dark, little sunlight penetrates the dense canopy, undergrowth is therefore limited – dense growth if trees are cut down
- huge variety of animals occupy different levels of the forest, few live on the ground: apes, birds, reptiles (snakes), insects; rivers contain many species of fish and reptiles

Soils:

- red clay soils rich in iron and aluminium oxides
- large numbers of decomposers
- high temperatures and humidity cause rapid chemical weathering:
  - rapid breakdown of leaf litter, fast recycling of nutrients through dense root systems; soils keep few nutrients and are infertile; nutrients are stored in the biomass
  - rapid weathering of parent rock gives very deep soils – up to **20 metres**
- soils have a weak structure and erode easily if exposed to heavy rain

Human activity:

- **shifting cultivation** (slash and burn) – traditional form of agriculture followed by local (**indigenous**) farmers (see also page 147)
- farming is supplemented by hunting, fishing and gathering (different fruit available all year round)
- extensive areas of forest are required to support a small number of families
- in many areas the system is under pressure from:
  - population growth: cleared areas are not given sufficient time to recover before being used again (**fallow period** is reduced) – soil erosion more likely and yields decline
  - large areas of forest are being cut down for logging and other new developments, reducing the forest available for traditional farmers

Fig 6.12 Flow diagram of shifting cultivation in the rainforest

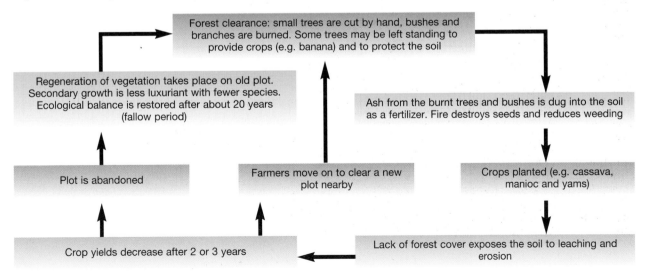

Forest clearance: small trees are cut by hand, bushes and branches are burned. Some trees may be left standing to provide crops (e.g. banana) and to protect the soil

Regeneration of vegetation takes place on old plot. Secondary growth is less luxuriant with fewer species. Ecological balance is restored after about 20 years (fallow period)

Ash from the burnt trees and bushes is dug into the soil as a fertilizer. Fire destroys seeds and reduces weeding

Plot is abandoned

Farmers move on to clear a new plot nearby

Crops planted (e.g. cassava, manioc and yams)

Crop yields decrease after 2 or 3 years

Lack of forest cover exposes the soil to leaching and erosion

Deforestation of tropical rainforests

> **KEY POINT** Deforestation is the cutting down and removal of areas of forest, in any part of the globe.

- Fifty per cent of tropical rainforest has been destroyed in the last 100 years.
- The rate of destruction has been increasing rapidly in the last twenty years.
- Some scientists estimate that if the current rate of clearance continues, most forests will have disappeared in the next ten years.

Causes of deforestation

- Extension of commercial agriculture – forest is cleared for cattle ranches.
- Timber extraction – hardwood for MEDCs (e.g. Japan and UK).
- Mining operations.
- Developing electricity supplies – HEP projects (reliable high rainfall).
- Road building – new roads allow easier access in and out the forest.
- Population growth:
  - within forest communities (shifting cultivators)
  - government resettlement schemes of urban poor, alongside the new roads
  - exploitative recreation and tourism – new hotels and leisure complexes.

Fig 6.13 Causes of deforestation in the Brazilian rainforest

This topic frequently appears in exam questions.

Benefits of deforestation

- Most rainforest is located in LEDCs (e.g. Brazil, Indonesia):
  - money earned from export of timber and ores will increase the GDP and lead to economic and social development and improve quality of life (schools, hospitals)
  - LEDCs require income to pay off loans from MEDCs.
- Jobs are provided by mining, logging and tourism.
- Resettlement provides better quality of life for people living in urban 'shanty towns'.

Problems of deforestation

- Nutrients are stored in the biomass. If trees are cut down, an infertile soil is left. New farmers will only obtain satisfactory yields for three or four years. After that, they will need to cut down more forest or abandon the farm.
- Exposed soils are quickly eroded by heavy rainfall. Clay soil is washed into rivers which causes deposition and flooding, making navigation difficult, ruining fishing and polluting water supplies.
- Rapid run-off of heavy rainfall on exposed areas causes flash floods and bank erosion.
- Indigenous people are losing their livelihood, homes and culture. They fall victim to diseases introduced by settlers which cause high death rates.
- Vaccines for diseases such as malaria have been developed from rainforest plants. Deforestation may prevent the discovery of new plants with possible medical value.
- Deforestation may be affecting climate change:
  - trees absorb CO_2 - loss of trees will increase the concentrations of CO_2 (greenhouse gas) in the atmosphere - a possible cause of global warming
  - burning trees releases more CO_2 into the atmosphere

Note the conflict between the need for LEDCs to develop and improve the quality of life of their people and global environmental considerations.

Conflicts in the use of tropical rainforest

> **KEY POINT**
>
> MEDCs suggest that LEDCs stop deforestation; accusing them of increasing global warming. But MEDCs demand most timber and are the largest importers. MEDCs burn 80% of the world's fossil fuels, which are the principle cause of global warming.

- LEDCs need the profits from the export of timber and minerals from their countries to improve their standards of living. MEDCs have already deforested the temperate forests and exhausted most of their mineral deposits.
- Locally indigenous people may be involved in violent conflicts with new settlers and loggers.
- Environmental groups such as the **World Wide Fund for Nature** and **Friends of the Earth** campaign to make the public aware and to persuade industry and governments to respond to the issues.

Solutions

- Protect **core areas** of forest free from any human activity (except for bio/medical research?)
- Outside core areas, establish a '**buffer zone**' and encourage economic but sustainable use of the forest by:
  - extractive reserves for rubber and gathering of fruit/nuts only

- timber to be felled in narrow strips and then allowed to regenerate
- small arable areas, protected by shading trees (e.g. banana palm).
- **Forest Stewardship Council** – links timber producers to customers and gives assurances that timber and timber products are only from areas of sustainable logging.
- MEDCs contribute to share the 'cost' of conserving the rainforest, and ensure that the living standards of all people in LEDCs improve.

1. What is an ecosystem?
2. By what process do green plants absorb light energy from the sun?
3. In a food chain, what is the name given to herbivores such as rabbits?
4. What is the principal source of nutrients in an ecosystem?
5. What name is given to plants which first colonise an area?
6. What name is given to global-scale ecosystems such as tropical rainforests?
7. What type of soil is found in coniferous forests?
8. What are the dominant trees in a temperate deciduous forest?
9. What types of vegetation are found in the savanna?
10. Where are tropical rainforests found?

1. Relationships between climate, soils and living organisms in an area 2. Photosynthesis 3. Primary consumers 4. Weathered rock 5. Pioneer species 6. Biomes 7. Podsol 8. Oak, ash, beech, elm 9. Drought-resistant trees and grasses 10. 5°N to 5°S of the Equator

Tectonic activity

After studying this section you should understand:

- that the crust of the Earth is divided into plates, which can be divided into continental and oceanic plates
- that tectonic activity results in earthquakes, volcanoes and fold mountains
- that most tectonic activity is associated with the plate margins
- that plate margins can be divided into three main types depending on how the plates are moving in relation to each other
- that each type of plate margin is linked to different forms of tectonic activity
- the main characteristics of earthquakes and volcanoes
- the hazards and benefits of volcanic activity
- the impact of earthquakes and volcanic activity in an LEDC and an MEDC
- the management issues (such as prediction and preparedness) associated with reducing the impact of earthquake and volcanic activity

LEARNING SUMMARY

Plate tectonics

AQA A   AQA C
EDEXCEL A   EDEXCEL B
OCR A   OCR C
NICCEA

To understand the location, distribution and causes of volcanoes and earthquakes, it is necessary to study **plate tectonics**.

KEY POINT   Plate tectonics is the theory explaining the movement, formation and destruction of the plates which make up the Earth's crust.

The crust

The **crust** forms the **outer surface** of the Earth. The crust:
- is cool enough to behave as a more or less rigid shell
- is up to 90 km thick (compared with a total radius of the Earth of 6400 km)
- floats on hotter, semi-molten rock (the **mantle**) below
- is sub-divided into plates:
  - seven large plates e.g. Pacific plate, Eurasian plate
  - twelve smaller plates e.g. Philippine plate, Juan de Fuca plate

Plates

There are two types of plate:
- **oceanic**
- **continental**

Fig 7.1 The internal structure of the Earth   **Fig 7.2** Principal crustal plates and plate margins

> **200 million years ago all the land masses of the Earth formed one continent known as Pangaea. Note how South America closely fits the shape of Africa.**

Oceanic plates are:
- composed of basaltic rock
- 5-10km thick
- denser than continental plates
- frequently destroyed beneath the continental plates
- younger than continental plates.

Continental plates are:
- mainly composed of granite
- 25-90km thick
- less dense and therefore can 'float' on the top of oceanic plates.

The plates are slowly moved by convection currents in the mantle.

Fig 7.3 How crustal plates move

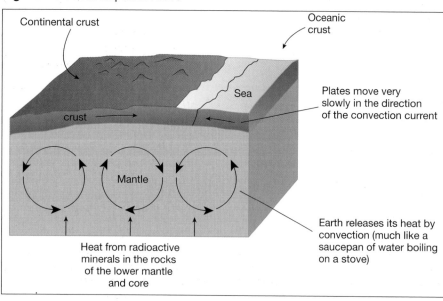

Plate margins

Where two plates meet is called a **plate margin**. Plates move in different directions and at different speeds, which causes stress and friction along the plate margins. Stress and friction are mostly responsible for:
- earthquakes and volcanic activity
- the creation of fold mountains (e.g. the Himalayas).

It is important to understand the difference between these margins. Know examples of where they occur, and relate them to your case studies of volcanoes and earthquakes.

Plate margins can be divided into three types (see Fig 7.2):

1 Destructive margins (or compressional or convergent margins)
   i subduction zones   ii collision zones
2 Constructive margins (or extensional or divergent margins)
3 Conservative margins (or tensional or transform margins)

Table 7.1 Types of plate margins

Type of margin	Description of movement	Tectonic activity	Examples (see Figs 7.8 and 7.9)
1 Destructive i Subduction zones (Fig 7.4)	• Ocean and continental plates converge • Denser oceanic plate sinks beneath the continental plate forming an **ocean trench** and **subduction zone** • Friction between the plates, and heat from the mantle, melts the oceanic plate which becomes molten magma • Under pressure, magma rises through the crust and forms a volcano • Friction and pressure between plates causes earthquakes • Pressure pushes the rocks upwards to form **fold mountains** • Where volcanic activity and fold mountains form off shore, an **island arc** is created	Large, violent earthquakes Violent volcanic activity Some of the largest volcanoes in the world Japan has 60 active volcanoes and frequent earthquakes	Nazca/ South American plates Japan Philippine /Eurasian plates Caribbean islands Caribbean / North American plates
ii Collision zones (Fig 7.5)	• Two continental plates collide • Equal density; neither plate sinks • Rock is compressed and forced upwards to form high fold mountains	Many earthquakes; sometimes violent	Himalayas Indo-Australian plate / Eurasian plate Alps Andes Zagros Mountains (south Iran)
2 Constructive (Fig 7.6)	• Plates move apart • Molten rock (magma) rises to the surface to fill the gap • Cools and forms **new** oceanic crust • Forms mid-oceanic ridges and volcanoes on the ocean floor • If a ridge or volcano rises high enough it will emerge at the surface to form a new island	Gentle volcanic activity Weak earthquakes	Mid-Atlantic Ridge Islands of Iceland, Surtsey (1963), Azores
3 Conservative (Fig 7.7)	• Plates move past each other, in similar or opposite directions • No crust is created or destroyed • Pressure builds up over time and is released in a sudden jerk • Violent shock waves are released causing earthquakes	Frequent, sometimes violent, earthquakes	San Andreas Fault (California) Pacific / North American plates

Fig 7.4 Destructive margin – subduction zone (cause of the destructive earthquake at Kobe, Japan in 1995)

Epicentre

Deep ocean trench
(subduction zone)

Japan
(Honshu)

Kobe
city

Pacific
Ocean

Eurasian Plate
(continental
crust)

Philippine Plate
(Oceanic crust)

Molten magma forced
upwards forming
the active volcanoes
found in many parts of Japan

Mantle

Pressure and friction
melts the oceanic crust
creating molten magma

Earthquake
focus - when
pressure is suddenly
released

Heavier oceanic plate forced
down beneath the lighter
continental plate

Fig 7.5 Destructive margin –
collision zone (Himalayas)

Fold mountains -
Himalayas growing
by several cms/year

Frequent earthquakes
in W. China

Latur earthquake
1993

Eurasian Plate
(moving south)

Indian Plate
(moving north and east)

Collision
zone

Mantle

Mantle

Rocks between the plates pushed
upwards, foldedand contorted

Fig 7.6 Constructive plate margin (Mid-Atlantic Ridge)

Mid-Atlantic Ridge
zone of volcanoes and earthquakes
mostly on the ocean floor

Surtsey

Iceland (volcanic island)

North
American Plate

Eurasian Plate

Mantle
As the plates move apart molten rock
rises from the mantle to form new crust

Fig 7.7 Conservative plate margin (San Andreas fault, California)

San Andreas fault line

Pacific Plate
moving north-west
at 4cms per year

North American
Plate
moving north-west
at 1cm per year

**Use labelled diagrams
such as these in your
revision and do not be
afraid of using them in
an exam. They are
quick to draw, and
show your knowledge
and understanding
clearly.**

San Andreas
fault line zone of friction and many earthquakes
e.g. San Francisco 1906, Los Angeles 1971, San Francisco 1989

Where are earthquakes and volcanoes located?

There is a clear pattern for the distribution of both earthquakes and volcanoes. The patterns are closely linked to the distribution of plate margins (Fig 7.2).

Questions aften ask you to describe the 'pattern' shown on a map. These two maps are good for practising your technique.

Volcanoes occur:

- in narrow bands; a pattern which is similar but not identical to earthquakes distribution, e.g. volcanoes are not found along collision margins (Himalayas)
- most often at destructive plate margins, e.g. around the Pacific Ocean – the 'Ring of Fire'
- along constructive plate margins, e.g. Mid-Atlantic Ridge
- at 'hot spots' in the centre of plates where the crust is thin, e.g. Mauna Loa, Hawaii.

Fig 7.8 Global distribution of volcanoes

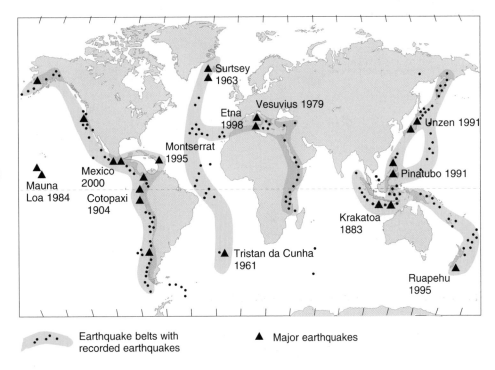

Fig 7.9 Global distribution of earthquakes

Earthquakes occur:

- in long narrow bands
- along all types of plate margin
- close to volcanoes
- on land or under water
- in a dense ring around the Pacific Ocean.

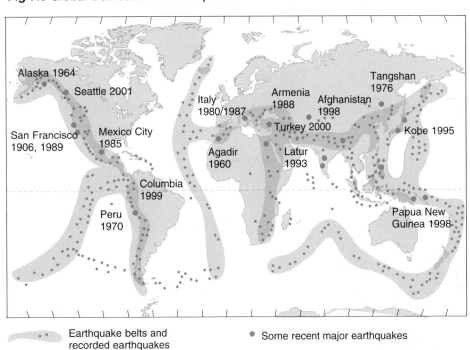

Volcanoes

AQA A AQA C
EDEXCEL A EDEXCEL C
OCR A
NICCEA

KEY POINT Volcanoes occur where weaknesses in the Earth's crust allow magma, ash, gas and water to erupt onto the land and seabed.

Fig 7.10 Characteristics of a volcano (Soufriere Hills volcano, Montserrat in the Caribbean)

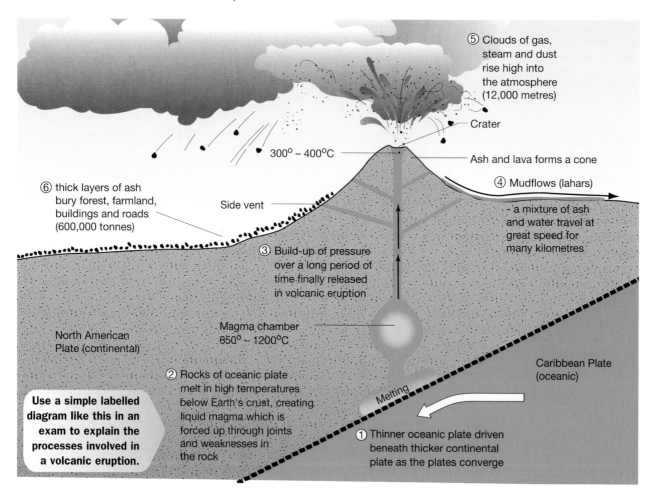

⑤ Clouds of gas, steam and dust rise high into the atmosphere (12,000 metres)

Crater

300° – 400°C

Ash and lava forms a cone

④ Mudflows (lahars)

⑥ thick layers of ash bury forest, farmland, buildings and roads (600,000 tonnes)

Side vent

– a mixture of ash and water travel at great speed for many kilometres

③ Build-up of pressure over a long period of time finally released in volcanic eruption

North American Plate (continental)

Magma chamber 650° – 1200°C

Caribbean Plate (oceanic)

Use a simple labelled diagram like this in an exam to explain the processes involved in a volcanic eruption.

② Rocks of oceanic plate melt in high temperatures below Earth's crust, creating liquid magma which is forced up through joints and weaknesses in the rock

Melting

① Thinner oceanic plate driven beneath thicker continental plate as the plates converge

Types of volcano

Most volcanoes occur at plate margins.

Volcanoes can also be classified in the following way (locations on Fig 7.8):

- **Active**: volcanoes that have erupted recently, e.g. Soufriere Hills, Montserrat; Mount St Helens, north west USA.

- **Dormant**: volcanoes that have not erupted for a long time, e.g. Mount Rainier, north west USA. Soufriere Hills volcano in Montserrat had been dormant for 400 years before the violent eruption in 1996.

Dormant volcanoes always pose an element of risk for the people living and farming nearby.

- **Extinct**: volcanoes that are no longer areas of tectonic activity; no record of volcanic activity since records began, e.g. Castle Rock, Edinburgh.

Fig 7.11 Types of volcano

Margin type	Destructive (subduction)	Hot spot	Constructive	Destructive (subduction)
Volcano type	Composite (ash and lava)	Shield	Usually shield	Composite (ash and lava)
Eruption	Explosive	Gentle oozing	Usually gentle	Explosive
Products	Nuées, ash, lava	Lava	Lava, ash	Nuées, ash, lava
Margin/plate	Caribbean/South American	Pacific	North American/Eurasian	Juan de Fuca/North American
Volcano	Soufriere, Montserrat	Mauna Loa, Hawaii	Surtsey, Iceland	Mount St. Helens, USA

Formation	At a destructive margin, one plate dives beneath the other. Friction causes it to melt and become molten magma. The magma forces its way up to the side of the actual plate margin (the ocean trench). A number of volcanoes may reach the surface to form a string of islands called an island arc, e.g. Japan.	In places where a plate is particularly thin, magma may be able to escape to the surface. Such a place is called a 'hot spot'. A shield volcano, like those on Hawaii, will be formed.	At a constructive margin where two plates are moving apart, new magma can reach the surface through the gap. Volcanoes forming along this crack create a submarine mountain range called an ocean ridge, e.g. the Mid-Atlantic Ridge.	Sometimes destructive margins involve continents. Instead of forming islands, volcanoes occur within a range of mountains, e.g. the Rocky Mountains and the Andes.

Hazards resulting from volcanoes

Table 7.2

Hazard	Description	Examples
Lava flows	Molten rock (magma) flowing down the sides of a volcano. Very hot basaltic lava flows very fast.	Frequent rapid lava flows on Mt Etna, Sicily, threaten local settlements.
Lahars	Mudflows, a mixture of ash and water, travel at great speed down the mountain sides.	Ruiz volcano, Columbia, erupted in 1985. The eruption melted snow around the summit and a mixture of water, ash and rock turned to mud and buried the town of Armero. 22 000 people drowned in the mud.
Dust and ash clouds	Ash thrown high into the atmosphere shuts out the sun and causes gloomy days. On settling, ash can completely bury buildings and crops.	When Soufriere Hills volcano in Montserrat erupted in 1996, ash clouds deposited 600 000 tonnes of ash burying the landscape and farms.
Nuée ardente (glowing cloud)	Flows of superheated ash and gases, flowing at very fast speeds.	In 1980, nuée ardente from the eruption of Mt St Helens in north west USA destroyed every tree over a very large area.

Benefits from volcanoes

> These benefits help to explain why large numbers of people live in the danger areas surrounding volcanoes.

- Volcanic ash weathers to produce very fertile soils, excellent for intensive farming, e.g. lower slopes of Mt Etna, Sicily.
- Geothermal energy – Iceland uses energy from volcanic water and steam to supply heat and electricity.
- Volcanic rock makes good building stone.
- Volcanoes attract adventurous tourists to visit sites in Iceland, New Zealand and southern Italy (Vesuvius and Stromboli).

Predicting volcanoes

Careful monitoring has allowed scientists to become more proficient at predicting volcanic eruptions. Early evacuation can be advised and lives saved. The main methods include:

- earthquake activity increases prior to eruption as the magma rise in the crust
- gas emissions change with increasing sulphur
- lasers and tiltmeters measure changes in the shape of the volcano; bulges and domes suggest a build up of magma
- ground temperature changes can be measured by satellite
- ultrasound is used to detect movements of magma
- studies of past activity may provide evidence.

Responding to a volcanic hazard

Although little can be done to control volcanoes, some measures that can be taken are:

- lava flows have been diverted from vulnerable areas
- lava flows have been halted by water sprays
- design buildings to shed ash
- communities can prepare for a disaster:
  - prevent settlement in vulnerable areas
  - organise evacuation procedures
  - improve public awareness.

LEDC case study: Montserrat, Caribbean, (British Territory) 1995-97

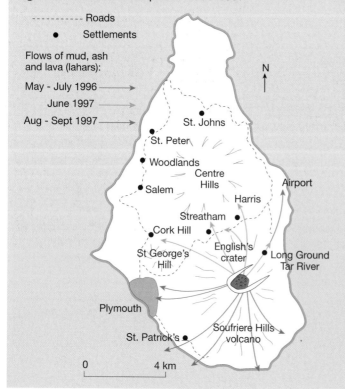

Fig 7.12 Montserrat: eruptions 1995-1997

- - - - - - - Roads
- Settlements

Flows of mud, ash and lava (lahars):

May - July 1996 ⟶
June 1997 ⟶
Aug - Sept 1997 ⟶

N ↑

St. Johns
St. Peter
Woodlands
Centre Hills
Salem
Harris
Airport
Streatham
Cork Hill
English's crater
St George's Hill
Long Ground
Tar River
Plymouth
St. Patrick's
Soufriere Hills volcano

0   4 km

Montserrat factfile:

Population: 11 000

An island in the island arc on the destructive plate margin between the North American and Caribbean plates.

Effects of the eruption

- Forests and rich farmland destroyed by lahars and covered with deep layers of ash.
- Many villages buried in ash; 23 people killed.
- Many people evacuated:
  - to the north of the island
  - to neighbouring islands
  - to the UK.
- The capital city, Plymouth, (port, industrial and administrative centre) almost totally destroyed and abandoned.
- The only hospital was destroyed.

- The only airport was burned and engulfed in ash.
- Many roads were destroyed.
- Dust covered the whole island, making it difficult to breathe.

Response to the eruption

- The authorities and people were totally unprepared since the volcano had been dormant for 400 years.
- People evacuated to the north of the island, were housed in tents and makeshift homes with little food, poor sanitation and no power.
- The UK government gave £55 million in compensation and redevelopment and 250 prefabricated houses.
- The hospital reopened in a former school.
- Communications were difficult and expensive to repair.
- Many people are still unemployed since the tourist industry has collapsed.
- Little farmland reclaimed in the south
- Little redevelopment in Plymouth; most people are resettling in villages further north while minor eruptions are still continuing.
- Considerable aid was given by the UK government to aid recovery.

MEDC case study: **Mt St Helens, north-west USA, 1980**

Fig 7.13 Effects of Mt St Helens eruption

- - - - → flows of superheated ash or gas (nuée ardente)

mudflows of ash and water (lahars)

• settlements

roads

flooded areas

Road and rail bridges washed away

Fish hatchery destroyed

Tourist and logging camps destroyed

All trees destroyed

Toutle

Trees and ash block rivers - widespread flooding

Avalanche

Zone of total destruction

Popular trout and salmon fishing rivers ruined

Ash deposits

Crater

Longview

N

Sediment carried downstream ruins barge transport on Columbia River

0   10
kms

(Continued on page 100)

Mt St Helens case study continued

Compare the response of the USA (an MEDC) with that of Montserrat (an LEDC) to an earthquake event.

Mt St Helens factfile

Mt St Helens lies on the destructive margin where the Juan de Fuca plate (oceanic) is disappearing below the North American plate.

Effects of the eruption

- Extensive forests and logging camps destroyed by nuée ardente.
- Ash and flooding destroyed crops and livestock.
- Lava flows and ash clogged prime salmon and trout rivers.
- Early warnings were given as scientists monitored the volcano; most people evacuated to safety.
- 63 people were killed, most from poisonous gases.
- Flooding washed away road and railway bridges.

Response to the eruption

- Ten million trees were replanted.
- Agriculture benefited from fertile ash.
- Rivers were dredged and restocked with fish.
- People were re-housed.
- Bridges were rapidly rebuilt.
- Tourism has increased, attracted by the event. A new tourist centre has been built, attracting a million visitors a year. New tourist lodges have also been put up.

Earthquakes

 AQA A   AQA C
EDEXCEL A   EDEXCEL B
OCR A   OCR C
WJEC

 KEY POINT

An earthquake is the vibration of the Earth's crust caused by shock waves travelling outwards from a sudden movement deep within the crust.

The source of the shock waves is known as the **focus**, and the point on the Earth's surface immediately above is the **epicentre**. Most earthquakes are associated with movements along plate margins, but many occur at other weaknesses in the crust such as fault-lines.

Fig 7.14 Characteristics of an earthquake

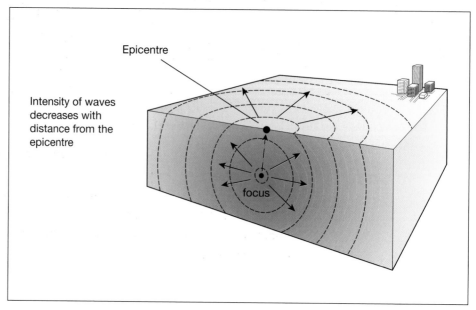

Epicentre

Intensity of waves decreases with distance from the epicentre

focus

Measuring earthquakes

The shock waves of an earthquake are recorded and measured on a **seismograph**. The seismograph measures the strength or magnitude on the **Richter Scale** (see Fig 7.15).

Fig 7.15 The Richter Scale

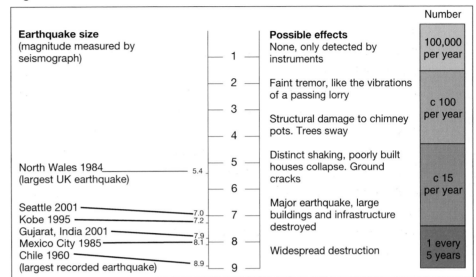

> Add examples of more recent earthquakes to Fig 7.15.

The impact of an earthquake

> 'The greater the strength of an earthquake, the ligher the death toll'. Is this statement correct? Use Table 7.3.

- Primary impact includes:
  - deaths
  - collapse of buildings
  - damage to roads, water and gas.
- Secondary impact includes:
  - fire and flooding
  - people made homeless
  - injuries and disease
  - loss of business

Table 7.3 The impact of individual earthquakes

Year	Place (see Fig 7.9)	Richter Scale	LEDC/MEDC	Impact
1960	Agadir, Morocco	5.7	LECD	12 000 deaths, many homeless
1976	Tangshan, China	7.8	LEDC	250 000 deaths, 650 000 homeless
1989	San Francisco, USA	7.1	MEDC	62 deaths
1994	Los Angeles, USA	6.7	MEDC	57 deaths, 20 000 homeless
1999	Turkey	7.3	LEDC	18 000 deaths, many remote villages destroyed
2001	Gujarat, India	7.9	LEDC	100 000 deaths, 30 000 homeless
2001	Seattle, USA	7.0	MEDC	1 death, 250 injuries, damage: billions of dollars

Factors affecting the impact of an earthquake

- The strength of the earthquake measured on the Richter Scale.
- The distance from the epicentre.
- The nature of the surface rocks – soft rocks can absorb water and become very mobile, e.g. Mexico City 1985.
- Population density: densely populated urban areas are particularly vulnerable, e.g. Kobe 1995.

- The remoteness of an area will affect the time that emergency services and aid take to reach the area, e.g. Afghanistan 1998.
- Resources available and preparedness of an area; MEDCs are able to afford better services.
- Time of day.
- Time of year/climate can affect survival and the spread of disease.
- Position with regard to impact of tsunamis.

> Tsunamis are not tidal waves. In Papua New Guinea (1997) a tsunami flooded seven coastal towns, causing 3 000 deaths and destroying many houses.

KEY POINT

Tsunamis (Japanese for 'harbour wave') are powerful ocean waves caused by earthquakes. They can travel long distances across oceans very quickly. As they approach the shallower water along a coast, the wave suddenly builds up to a great height. The speed, height and power of the wave can cause severe damage and many deaths in low-lying coastal areas.

Responding to earthquakes

Control

Research in the USA has experimented with injecting fluids along the San Andreas fault to lubricate zones of friction (see Fig 7.7). There is little evidence that this works.

> Despite much research, it is still not possible to predict earthquakes.

Prediction

It is possible to predict where earthquakes are likely to happen but not when. On-going research is investigating:

- evidence from past earthquakes
- patterns of small tremors
- unusual animal behaviour
- gas emissions from the ground.

> One of the reasons for the low death toll and injuries in Seattle (2001) was that large sums of money had been spent over the last 20 years to ensure buildings can withstand major shocks. The Space Needle, built to withstand a 9.1 magnitude earthquake, shuddered violently but no damage occured.

Preparation

Preparation for earthquakes and their aftermath can include:

- land use planning to prevent building on weak rock, e.g. reclaimed land from the sea
- improved building design, including:
  - rolling weights on roof to counteract the shock waves
  - identification number on roof visible to helicopters assessing the damage after an earthquake
  - 'bird cage' interlocking steel frames
  - automatic shutters that come down over the windows to prevent pedestrians being showered with glass
  - panels of marble and glass flexibly anchored into the steel superstructure
  - rubber shock absorbers between the foundations and superstructure
  - reinforced foundations deep in bedrock
  - open areas outside buildings where pedestrians can assemble if evacuated
- more flexible gas, water and power lines
- planned emergency and evacuation procedures
- greater public awareness – education in schools, e.g. Japan and California.

MEDC case studies: Kobe, Japan, 1995

Kobe factfile

Location: Kobe, Japan (see Fig 7.9); a heavily populated urban area

Time and date: 5.46 a.m. on 17th January 1995

Strength on Richter Scale: 7.2

Plate margin: destructive margin – subduction zone (see Fig 7.4)

Short-term effects

- 5477 deaths
- 35 000 injured
- 316 000 left homeless
- Many houses collapsed despite earthquake proof designs.
- Large number of buildings destroyed by fire when gas mains fractured.
- Many people evacuated to schools, parks and community centres; tents and emergency rations provided.
- Rescue teams arrived quickly on scene with equipment.
- National government set up emergency headquarters in two days.
- Injured people transferred to hospitals in nearby cities.
- Electricity restored in six days.
- Search for survivors concluded in ten days.
- Earthquake refugees start moving into temporary housing in two weeks.
- Much industry, communications and shopping restored within three months.
- All temporary shelters for homeless closed in seven months.

Long-term effects

- Many businesses closed.
- Extra jobs created in rebuilding.
- Many people moved away from area permanently.
- Overall cost 10 billion yen.

LEDC case study: Latur, Maharashtra, central India, 1993

Latur factfile

Location: Latur, Maharashtra, India (see Fig 7.2); a well-populated, remote, rural area

Time and date: 4.00 a.m. in September 1993

Strength on Richter Scale: 6.4

Plate margin: Destructive margin – collision zone (see Fig 7.5)

> Be prepared to explain why the effects are far more severe in LEDCs than in MEDCs.

Short-term effects

- 22 000 deaths
- 4800 injured
- 10 000 left homeless
- Nearly all the stone houses collapsed, bringing rock and concrete down onto sleeping people.
- Large numbers of people camped in makeshift shelters made from tin sheeting, straw and polythene.
- Army rescue teams did not arrive for 36 hours.
- Large numbers of dead posed threat of disease and polluted water supplies.
- Lack of hospital service; too remote to transfer to other towns.
- Rescue and retrieval of bodies continued for many weeks.
- Earthquake refugees still in camps with poor sanitation many months after the event.

Long-term effects

- Very little temporary housing was provided.
- Some villages remain in ruins.
- Farming, which relies on hand labour, was devastated by loss of life.
- Families were left destitute.
- International aid agencies were mainly responsible for rehabilitation.
- Indian government asked for aid to meet the £22 million costs of rescue.
- Long-term aid for the region was insufficient.

PROGRESS CHECK

1. What are the two types of plate forming the Earth's crust?
2. What type of plate boundary is causing the formation of the fold mountains of the Himalayas?
3. What type of plate boundary is found close to Japan, at the boundary between the Eurasian and Pacific Plates?
4. What type of plate boundary is found along the Mid-Atlantic Ridge?
5. What is the name given to the molten rock, which flows down the sides of some volcanoes?
6. How is tectonic activity of benefit to Iceland?
7. What is the name given to the point on the Earth's surface immediately above the focus of an earthquake?
8. What scale is used to measure the strength or magnitude of an earthquake?

1. Oceanic/continental 2. Destructive/collision zone 3. Destructive/subduction zone 4. Constructive 5. Lava 6. Tourism, geothermal energy 7. Epicentre 8. Richter Scale

Population

After studying this section you should be able to:

LEARNING SUMMARY

- describe and understand the main features of the world pattern of population density
- understand the main reasons for population change and the effect on population structure
- describe a range of population issues and their planned solutions

The world pattern of population density

AQA A   AQA B   AQA C
EDEXCEL A   EDEXCEL B
OCR A   OCR B   OCR C
WJEC A   WJEC B
NICCEA

KEY POINT

Population density is measured as the number of people living in an area. It is usually measured in people per square kilometre.

People are not evenly spread across the Earth's surface; see Table 8.1.

Table 8.1

Be sure to know some examples of places with different population densities.

Scale	Dense	Sparse
International	SE Asia, Western Europe	Sahara Desert, Antarctica
National	Singapore, Netherlands	Australia, Mongolia
Regional	SE England, SE Brazil	Western Australia, Amazonia
Local	Hong Kong, Rio de Janeiro	Snowdonia, Alaska

Fig 8.1 World population distribution

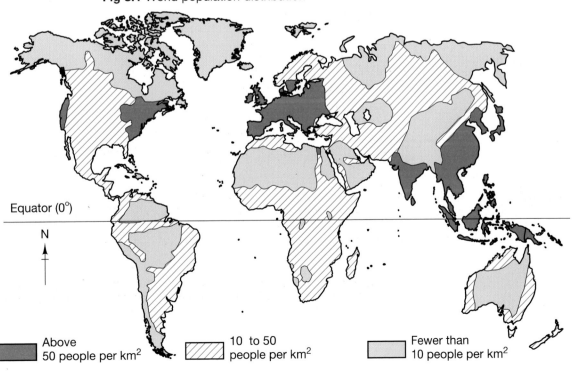

Equator (0°)

N

Key | Above 50 people per km² | 10 to 50 people per km² | Fewer than 10 people per km²

Reasons for the differences in population distribution and density

The factors affecting the distribution and density of population include:
- physical factors such as relief, climate, vegetation
- social factors such as the proportion of urban to rural population
- economic factors such as the resources, amount of industrialisation
- political factors such as government policy on birth control.

> **KEY POINT** Population distribution describes the pattern of people in an area (how the population is spaced or spread out). The three main patterns of distribution are uniform or even, nucleated or clustered, and a random pattern.

MEDC case study: population distribution in Japan

Fig 8.2 Population density in Japan

Japan factfile

Area: 380 000 km² (1.5 times the size of the UK)

Population density: 350 people per km² (1.5 times the density of the UK)

Description: Japan is made up of four islands. To the north is the sparsely populated Hokkaido. The largest island, Honshu, is densely populated and contains many large cities, including Tokyo. The northern part of Honshu, and the two southern islands of Shikoku and Kyushu, have intermediate population densities.

LEDC case study: population density in Brazil

Fig 8.3 Population density in Brazil

Brazil factfile

Area: 8 511 965 km²

Population density: Brazil has a population density of 20 people per km².

Description: Brazil can be divided into five regions. The North is sparsely populated and contains the Amazon river and a large area of tropical rainforest. The main city is Manaus. Population is increasing as economic development takes place.

The North East contains one in three Brazilians, but its population is declining because of drought.

The South East has the highest population density and includes the cities of Rio de Janeiro and São Paulo. This is the economic centre of Brazil. The South has a high population density with good agriculture and growing industry.

The Centre West area has a low population density but this is growing because the new capital city of Brasilia has been built there.

PROGRESS CHECK

1. Name two areas of high and two areas of low population density.
2. What is the difference between population density and population distribution?
3. What factors tend to account for areas of low population density?

1. Examples of high population density: London, Singapore, the Netherlands; low population density: Sahara Desert, Amazonia. 2. **Density** refers to the number of people per km²; **distribution** is the pattern of population such as nucleated or dispersed. 3. Rainfall, temperature, relief, access.

World population growth

OCR A   OCR B   OCR C
AQA A   AQA B   AQA C
EDEXCEL A   EDEXCEL B
WJEC A   WJEC B
NICCEA

The best estimate of the world's population total is six billion (one billion is equal to one thousand million). The total has doubled since 1960 with most of the growth in LEDCs. Projections of further growth vary. Some experts suggest that the total will reach eight billion. Others believe that social and economic changes in LEDCs will slow their population growth. Many countries take a census (questionnaire survey) every ten years to count their population. A census provides useful information for social and economic planning.

Fig 8.4 World population growth

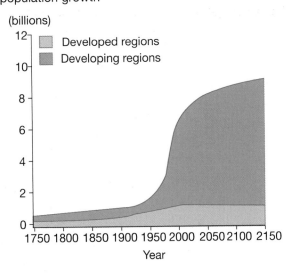

Population change: the balance of birth and death rates

KEY POINT   Birth rate is the number of live births per 1000 people per year.
Death rate is the number of deaths per 1000 people per year.

Know one case study of population change in an LEDC and one in an MEDC.

Natural population increase is the difference between **birth** and **death** rates. Natural increase will occur if births exceed deaths.

Birth rates are high in LEDCs such as Bangladesh and India because:

- children provide labour on family subsistence farms
- there are no old age pensions so children provide security for old age
- large families are seen as a sign of the husband's virility
- girls are expected to marry early thus extending their child-bearing years
- women are expected to stay at home and raise a family; they have little education and do not know about birth control
- some religions do not approve of contraception
- high infant mortality encourages large families to ensure some children survive.

Birth rates are low in MEDCs such as the UK, Sweden and Germany because:

- people marry later, women are educated and delay having a family so that they can follow a career
- high costs of living make children expensive
- couples prefer material possessions such as a car, house and holidays
- birth control, and the contraceptive pill in particular, are available
- governments discourage large families to save the costs of building more schools.

Death rates are low in MEDCs and are falling in LEDCs because:

- better healthcare and medical care are available
- people are able to retire earlier after less physically demanding jobs
- cures are being found for diseases such as cancer, malaria and cholera
- people are better educated about hygiene
- there are cleaner water supplies and sanitary facilities
- higher incomes buy the right food, housing and heating.

The Demographic Transition Model

Geographers studying the change taking place in the populations of countries have noticed that they follow a similar pattern. This is called the Demographic Transition Model. The model shows how the total population of a country changes through time as birth and death rates change.

The demographic transition model is useful for:

- studying the way population is changing
- understanding trends in births, deaths and natural increase
- predicting the changing structure of population and planning to meet its changing needs.

Fig 8.5 The Demographic Transition Model

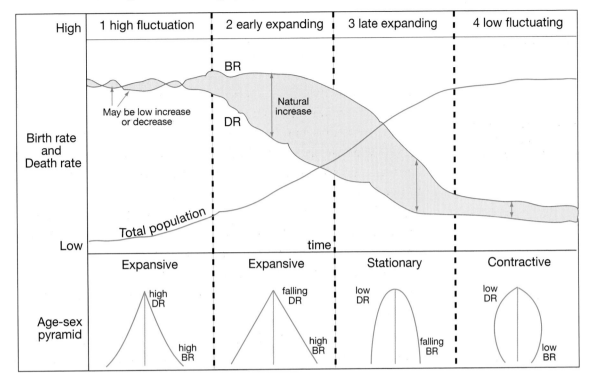

Stage	One	Two	Three	Four
Birth rate	High	High	Falling	Low and varies
Death rate	High and varies	Falls	Low	Low
Population changes	Small	Rapid increase	Slower growth	Stable
Places	Rainforests	Malawi	China	UK, Germany
People's lives	Subsistence agriculture; high infant mortality	Better food supply; control of disease; lack of birth control	Better living conditions and healthcare; growth of industry, jobs	High standards; affluence; small families; long life expectancy; education for all

This is a very popular diagram with examiners. Note how total population is changing at each stage.

The limitations of the Demographic Transition Model are:

● LEDCs may not follow the patterns of change found in MEDCs 30 to 50 years ago

● birth rates have not fallen as rapidly as might be expected in some LEDCs because of social customs and beliefs

● government planning for population change may interrupt the model, e.g. the 'One-Child Policy' in China

● some industrialising LEDCs are moving more rapidly through the stages than the MEDCs did.

Population structure

Divide the pyramid into three age groups.

The structure of a population can be studied using a population pyramid.

Fig 8.6 Population pyramid for France (MEDC)

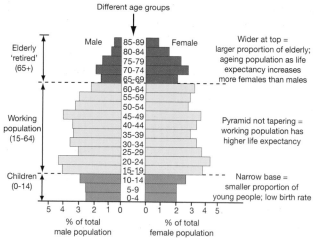

Fig 8.7 Population pyramid for Brazil (LEDC)

PROGRESS CHECK

1. Make a list of the reasons for high birth rates in LEDCs.
2. Why are death rates tending to fall in most countries?
3. Name the four stages of the Demographic Transition Model.

1. See page 108 2. See page 108 3. See page 109

Population change

AQA A AQA B AQA C
EDEXCEL A EDEXCEL B
OCR A OCR B OCR C
WJEC A WJEC B
NICCEA

Population migration

KEY POINT
A migrant is a person who moves to live in another place either permanently or temporarily.
Immigrants move into an area and emigrants move out.

Types of migration

Migration is not part of natural population increase.

There are four types of migration:

- rural to rural, e.g. a farmer retiring to a cottage near the sea
- rural to urban, e.g. people moving from the countryside to cities for jobs and other amenities
- urban to rural, e.g. families moving to a house with a garden in a village from where working members commute back to the urban area
- urban to urban, e.g. people moving from one urban area to another for a job change or moving from a big city to a smaller urban area for better housing.

Another division of migration is into forced and voluntary migrants.

- Forced migrants have had to move, e.g. Jews from the Nazi regime in Germany in 1939.
- Voluntary migrants move because they believe there are benefits if they move.

A model for understanding migration

Fig 8.8 The push–pull model of migration

Push factors	Journey factors	Pull factors
Lack of jobs	Cost of the journey	Job opportunities
Poor housing	Other attractions on the way	Housing
Poor environment e.g. pollution, crime, traffic, dust, eyesores	Running out of money for travel	Green areas, safety, privacy
	Poor transport	Good schools
War, genocide, persecution	Natural disasters e.g. floods	Affordable healthcare for all
Poor schools	Long distances	Using wealth to improve life
Lack of healthcare	Language difficulties	Slower pace of life
Retirement	Homesickness	Friends
Family breakdown		People of similar roots
Racial tension		New experiences
Restlessness		New community emerging
Decline of a community		Increased wealth

Case study of rural-urban migration in south-east Brazil

> Often fathers and older sons move first and the rest of the family follows later.

Rural to urban migration is common in many LEDCs. Many rural people in Brazil migrate to the cities in the south east of the country because:

- high population increase causes land and food shortages
- people believe the urban area is a place of 'bright lights' and opportunity
- crop failures, lack of capital, absentee landlords and lack of education mean a life of rural poverty
- farm mechanisation causes unemployment.

Case study of regional migration in Italy

> Note the way the push-pull model helps you to understand this migration.

The major movement is from the south, the Mezzogiorno, to the north of Italy. The main factors are:

- low standards of living based upon agricultural employment
- lack of educational opportunities
- steep slopes and poor communications make the movement of goods difficult
- lack of raw materials and poor power supplies
- poor agricultural development because of poor soils and lack of water in summer
- underdeveloped tourist industry so lack of jobs
- failed attempts by government to introduce new industry to the area
- attractions in the north include large cities, rich industry, water supply and better conditions for farmers.

Population problems and issues

> **KEY POINT**
> Overpopulation occurs when there are too many people for the available resources.
> Underpopulation occurs when there are too few people to fully exploit the available resources.

The problems of population growth in LEDCs

Many LEDCs have high rates of population growth because of high birth rates and declining death rates. The problems caused by this growth include:

- unemployment
- housing shortages
- lack of food
- shanty towns
- crime
- prostitution
- strain on services
- overcrowded buses
- poor sanitation
- poor medical care.

However, this growth provides a labour force and an increased domestic market.

> **KEY POINT**
> Population dependency is the ratio of people not economically active to those economically active. The dependent population includes age groups 0-16 and over 60 or 65. A dependency ratio of 1 means a dependent is supported by one member of the working population. LEDCs have dependency ratios greater than 1 because of the large numbers of young dependants. Some MEDCs also have dependency ratios over 1 because of the increasing number of elderly people.

> **KEY POINT**
> Life expectancy is the number of years that a person is expected to live at birth.

The problems of ageing populations in MEDCs

Life expectancy is increasing in many countries because of better diets, housing and healthcare as well as less physically demanding jobs, e.g. the 'greying' population in Sweden, Japan and the UK. Birth rates have also fallen in these countries. The problems caused by ageing populations include:

- increased funds needed to meet the demand for health services
- specially designed houses and serviced blocks of flats, e.g. hand rails, non-slip floor, emergency buttons
- provision of leisure activities for the elderly
- increased dependency and burden on the working population
- shortages of labour and the need to bring in workers such as nurses and doctors.

> A growing problem needing special provision; good for exam answers.

However elderly people do undertake many worthwhile tasks, many of them voluntary, as well as supporting their grown-up children as they raise their young families.

The problems of rural to urban migration in LEDCs

Migrants arrive in cities with little money and few skills for urban jobs. They cannot afford housing and are forced to settle in temporary dwellings made from scrap materials such as corrugated iron, wood, cardboard and cloth. These residential areas have:

- little or no sanitation and poor water supply
- unsafe electricity taken illegally from the National Grid
- few health facilities and widespread disease
- overcrowding, unemployment and crime.

The effect of this movement is that rural areas do not have enough labour; women with the help of the very young and the elderly often run small farms. As a result, farming output falls.

> **Know at least one example of source and destination.**

The problems of urban to rural migration in MEDCs

With improvements in public transport and increased car ownership, people are moving away from large urban areas to live in the smaller market towns and villages nearby. This causes problems for these country areas including:

- population increase places a strain on housing and health services
- increase in traffic creates congestion, noise and pollution
- schools may have to be expanded
- housing prices rise and local young people are not able to afford to buy in the area
- loss of open space as developers build houses to meet the demand
- public transport becomes crowded as commuting increases.

The effects on the urban area are not always positive, as the people left behind are often those with fewer skills and some social problems. The arrival of affluent people in a village can revitalise the area allowing shops to remain open, the village school to prosper and clubs and societies to flourish.

> **Know the effects on source and destination areas.**

The problems and issues raised by refugees

Some people are forced to leave their home area or country. They are called **refugees** and estimates suggest there are 30 million of them in the world. The reasons people become refugees include:

- persecution due to race, religion, nationality, social group or political beliefs, e.g. the Vietnamese 'boat people' in 1975 with the fall of South Vietnam; and the Muslims in Serbia due to the ethnic cleansing by the government
- civil war between rival ethnic groups such as that in the 1990s in the African country of Rwanda
- natural disasters such as earthquakes, volcanic eruptions, floods and droughts, e.g. Ethiopians leaving the Sahel region due to drought at a time of high population growth and poor harvests.

The problems and issues raised by these refugees include:

- they arrive in neighbouring countries that do not have the resources to cope with them, e.g. Ethiopian refugees filled camps in nearby Somalia which is another very poor country

- they usually arrive in large numbers in a very short time and have to be housed in tented camps with poor sanitary and water facilities; disease often breaks out
- they need food, water, medicine and shelter.

After the emergency some refugees return home. Others stay in their new country. They may bring valuable skills and expertise and find good employment. Others fill menial jobs that local people will not do. However, there are refugees who do not settle; they find the language, currency and customs of their new country difficult. For a sense of security they tend to stay together in the same neighbourhood and create a small piece of their own country. Racial conflict can occur.

Fig 8.9 Some examples of international migrations

(From Far East)

(To USA)

A

B

C

D

A temporary Mexican workers to USA (voluntary)
B permanent New Commonwealth to UK (voluntary)
C temporary Rwandan refugees to neighbouring countries (forced)
D permanent Vietnamese boat people to Pacific Rim countries (forced)

Strategies for coping with population change

Strategies for reducing population growth in LEDCs

Many LEDCs face increasing population totals and more mouths to feed. This takes resources away from plans to improve the **quality of life** of people. The strategies adopted by LEDCs include:

- **birth control** using contraceptives, abortion and sterilisation (tried in India with limited success)
- raising the level of education, especially for girls
- encouraging delayed marriages or raising the age at which people may marry
- making polygamy (multiple spouses) illegal
- raising the status of women to have a view on their number of children
- better health removes the need to have many children in case some die
- financial benefits to couples with small families
- providing employment opportunities and careers especially for women
- China has a **one-child policy** that encourages couples to have only one child. This policy includes:
  - birth control and fines if more than one child is born
  - increased taxes for couples with more than one child

Many candidates remember the 'one-child policy' but say it comes from Japan!

- public education and advertisements on the reasons for the policy
- salary bonuses of 10% for those with one child
- priority in education, housing and health for one-child families.

This harsh policy has been relaxed particularly in rural areas. It created a generation of 'spoiled, only child'. Some girl babies were abandoned or murdered because parents wanted a son to keep the family name alive.

> **KEY POINT**
>
> **Population strategy: plans by governments to solve the problems of population growth or decline.**

Strategies for increasing population in MEDCs

This has become an important topic so expect examination questions on it.

Some MEDCs have low birth rates, increasing numbers of elderly people, and a falling population total. The population is not at its **replacement level**. These countries face **labour shortages** and do not meet economic plans for improvements in the quality of life of the people living there. The strategies adopted by these MEDCs include:

- in Japan, they have raised the age of retirement to increase their workforce
- in Singapore the government has:
  - encouraged single people to marry
  - encouraged married couples to have more babies
  - provided a Social Development Unit that helps single people to meet each other
  - established child care centres for working women
  - introduced flexible working hours and part-time employment for mothers
  - extended the period of maternity leave
  - given tax rebates to married couples with more than two children
  - subsidised child care centres
  - given housing priority to couples with more than two children
  - advertised the benefits and joys of marriage on television programmes
  - shown stories of happy family life in newspapers.

Reducing the pressure of population growth

Planned migration

It is worth looking at an atlas to see where these places are found.

Indonesia has planned for the movement of some of its people from the crowded islands of Java (population density 700 persons per km²) and Bali to the less crowded islands of Sumatra, Kalimantan, Sulawesi and Irian Jaya (population density 3 persons per km²). This is called a transmigration policy. It is the world's largest resettlement scheme. This has:

- reduced overcrowding and unemployment in Java and Bali
- lowered the pressure on the land in Java and Bali
- eased rural to urban migration
- made use of large areas of forested land on the other islands.

Under the plan, land is given to poor farmers, together with some seeds, a house and food. It has been successful in Sumatra. Some migrants have found jobs on plantations opened on the hilly land. The migrants have more skills than the local people and there have been clashes. The migrants have challenged traditional culture and taken some urban jobs from local people.

Fig 8.10 Transmigration in Indonesia

There has been soil erosion in some of the forest areas that were cleared for cultivation. Java remains a very crowded island.

Strategies for coping with the elderly in MEDCs

Increased life expectancy means that governments and local authorities in North America, Western Europe and Japan have to make provision for the elderly. The types of provision they need to make include:

> There will be purpose-built housing for the elderly in a town near you.

- housing which is affordable and easily managed by older people
- health provision to meet the needs of the elderly including specialist cancer and heart units
- car parking and shopping facilities which are suited to people less steady on their feet
- access to social and religious services which will include more home visits
- library and other leisure facilities.

PROGRESS CHECK

1. What is the difference between over population and under population?
2. Define life expectancy and explain why it is increasing.
3. Name a group of refugees and explain why they wish to move to another country.

1. See page 112 2. See page 112 3. See pages 113–114

Settlement

LEARNING SUMMARY

After studying this section you should be able to:

● *describe the site and situation of rural and urban settlements*
● *recognise the functions of settlements and the services they offer*
● *know about urban land use and zones within settlements*

The location of settlements

AQA A AQA B
EDEXCEL A
OCR C
NICCEA

The location of settlements involves the study of both site and situation of different settlement types.

Different types of settlement

There are two types of settlement, **rural** (e.g. a village) and **urban** (e.g. a city). Residents in rural settlements are mainly farmers or workers from urban settlements. Urban settlements are involved in industry, commerce or administration. Rural settlements are smaller in size, population and population density. Urban settlements include towns, cities, conurbations and megalopolises.

KEY POINT

Rural: small number of buildings, agricultural or dormitory function, low density, close community, small number of services, in countryside.
Urban: continuous built-up area, with industry and commerce, high population density, acquaintances rather than friends, many services.

The siting and situation of settlements

KEY POINT

Site: the features of the place where the settlement is located.
Situation: the features of the area in which the settlement functions.

The factors influencing the **site and situation** of settlements include:

● dry point in an area of poor drainage, e.g. Ely, and wet point where water supply is good, e.g. Gretton, Gloucestershire
● spring line, e.g. Princes Risborough
● shelter and defence, e.g. Durham, Edinburgh, Conwy
● resources such as minerals, e.g. coal
● communications: gap towns, e.g. Dorking, route centres, e.g. Crewe, and bridging points, e.g. Worcester
● planned: new towns, e.g. Harlow, expanded towns, e.g. Swindon
● resorts, e.g. Bournemouth, spas, e.g. Buxton

Check whether a question asks for a site or situation. If it asks for location it means both.

- ports, e.g. Grimsby, Southampton
- religious centres, e.g. Canterbury
- market centre, e.g. Norwich

These factors were important when the settlement was founded but are less so today.

Fig 9.1 A spring village line

Settlement patterns

There are two questions to ask when looking at settlement patterns.

- Are the settlements in a nucleated, random or uniform pattern (Fig 9.2)?
- Do the settlements have a linear, nucleated or dispersed layout (Fig 9.3)?

Fig 9.2 Nucleated, random and uniform settlements

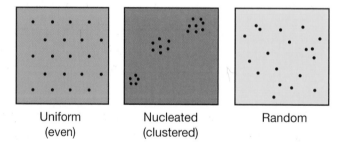

Uniform (even)   Nucleated (clustered)   Random

Fig 9.3 Linear, nucleated and planned dispersed layouts

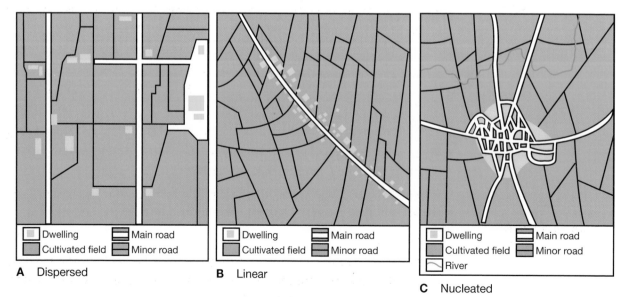

A Dispersed

B Linear

C Nucleated

The function of settlements

AQA A   AQA C
OCR A   OCR B

KEY POINT Settlement function: describes the main economic activity of the settlement, the jobs of the people working there.

Settlements can be **classified** according to their main function:

- manufacturing, e.g. Sheffield
- hi-tech industry, e.g. Reading
- holiday resort, e.g. Blackpool
- pottery, e.g. Stoke on Trent
- educational, e.g. Oxford and Cambridge

All larger settlements also act as centres for administration of the area around them.

> **Know the names and functions of towns in your area.**

Services in settlements (includes hierarchy, range of good, threshold, sphere of influence)

Settlements provide housing, administration, industry, commerce and services. The services include local government offices, shopping centres, schools and hospitals. Settlements can be arranged into a hierarchy according to their population size, range of services and distance apart.

Central Place Theory attempts to explain the size and spacing of settlements and the services they offer using the following ideas.

Goods

Goods purchased regularly, such as bread and newspapers, are called low-order or convenience goods. They are usually available in the village or corner shop.

Goods purchased irregularly are called high-order or comparison goods, e.g. television, clothing, furniture. People compare prices on these goods; they want choice and are prepared to travel further to city centres or retail parks.

Range of a good

This is the maximum distance that people are willing to travel for a service. They travel further for high-order, comparison goods.

Threshold population

This is the minimum number of customers needed to maintain a service. Small local shops need a small threshold population buying low-order goods regularly. Comparison goods shops need a larger threshold population because goods are bought infrequently. Very specialist shops, e.g. jewellers, need a very high threshold population and are usually only found in large settlements.

Sphere of influence

This is the area served by the settlement. It is the area from which people travel to use the settlement or the area to which bus services run and deliveries are made. Newsagents have small spheres of influence, secondary schools have larger spheres than primary schools and hospitals have very large spheres of influence.

> **Be able to apply range, threshold and sphere of influence to two settlements you know.**

Hierarchy

The hierarchy of settlements extends from metropolitan areas, through cities, towns and villages to hamlets. The number of services offered and the sphere of influence increase up this hierarchy.

Fig 9.4a Hierarchy of settlements

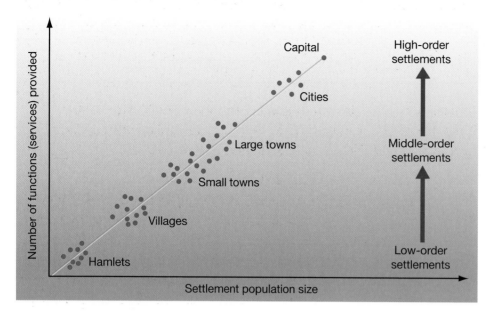

Fig 9.4a shows there is a large number of low-order settlements and a small number of high order settlements.

Fig 9.4b Hierarchy of shops and services

Fig 9.4b shows the pattern of movement of people to settlements for (i) food, (ii) solicitors and (iii) hospitals.

Fig 9.4c Sphere of influence

Fig 9.4c shows the spheres of influence of a town for three different services.

PROGRESS CHECK

1. What is the difference between the site and situation of a settlement?
2. Name an example of each of the following:
   ● a wet point settlement ● a resort ● a new town ● a market town.
3. Which of the following has the largest sphere of influence:
   ● a newsagent ● a supermarket ● a furniture shop ● a public house?

1. See page 117 2. See pages 117–118 3. See page 119, a furniture shop

Models of land use in settlements

AQA A AQA B
EDEXCEL A
OCR A OCR C
WJEC A
NICCEA

KEY POINT

Surveys of land use in settlements show areas with similar land use. These are called land use zones. For example, a zone of lower-quality terraced housing can be found near to the CBD.

To simplify and explain the complex pattern of land use in settlements a number of models have been developed. The three main models are the **concentric zone**, the **sector** and the **multiple nuclei**.

Fig 9.5 The concentric zone model

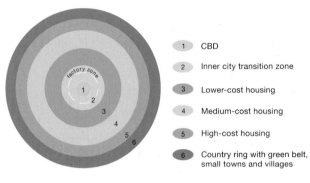

1 CBD
2 Inner city transition zone
3 Lower-cost housing
4 Medium-cost housing
5 High-cost housing
6 Country ring with green belt, small towns and villages

Fig 9.6 The sector model

1 CBD
2 Inner city transition zone
3 Lower-cost housing
4 Medium-cost housing
5 High-cost housing
6 Country ring with green belt, small towns and villages

Fig. 9.7 The multiple nuclei model

1 CBD
2 Zone in transition/ light manufacturing
3 Low-class residential
4 Medium-class residential
5 High-class residential
6 Heavy manufacturing
7 Outlying business district
8 Residential suburb
9 Industrial suburb
10 Commuter zone

Think of ways to find the edge of the CBD.

The concentric model

This model was based upon studies of American cities and particularly on Chicago in the 1920s. It has the Central Business District (CBD) at its centre. This zone is mainly made up of shops and offices. Around the CBD is the zone of transition or twilight zone or inner city. This is an area of older housing, declining industry and some derelict land. Outside this are three housing zones, which increase in quality away from the centre.

The sector model

In the sector model lines of transport to the centre are given more emphasis. Industry (in zone 2) and lower-cost housing developed along the line of the main roads or railways.

The multiple nuclei model

The multiple nuclei model shows different land uses clustered together and there may be several centres which are nearly as important as the CBD.

No settlement fits these models perfectly. However, many settlements do shows aspects of these models very clearly.

Land use in the Central Business District

Competition for land makes land prices high in the CBD. The CBD contains the main retail and commercial premises, covered shopping arcades, pedestrianized areas, major public buildings and administrative headquarters. Shops sell high-order goods and high-rise buildings are present. There is a high daytime population but few people live there. Land use and buildings are continually changing in the CBD. The congested CBD faces competition from out-of-town centres. This area is called '**downtown**' in the United States.

Land use in the inner city

The land around the CBD is called 'the inner city', the 'twilight zone' or 'zone of transition'. In this zone there are derelict old factories and poor terraced housing, built in the nineteenth century. Some of these 'brownfield' sites are being replaced by CBD buildings, large warehouses or new housing, roads and railways. During the period of redevelopment they become temporary car parks. Many people living in the inner city suffer deprivation (unemployment, overcrowding, crime, and pollution). Government policies have tried to redevelop, renew, rehabilitate and gentrify these problem areas, e.g. Hulme in Manchester, Toxteth in Liverpool.

> **KEY POINT**
> Greenfield sites are open land that has not been built on before. Brownfield sites are vacant or under-used land that has been built on and is ready for redevelopment.

Land use in the suburbs

Away from the CBD land prices fall. Houses and gardens become bigger and people living there can use their car or public transport to get to work or the city centre. As population and mobility increased, cities sprawled over the surrounding countryside. Factories and offices have moved from the inner urban area to 'greenfield' sites in the suburbs near to motorways or ring roads. Many local authorities moved people from the inner areas to large council estates in the suburbs. Some of these have become problem areas. People living in the suburbs tend to segregate into areas of similar class, income or ethnic background.

The country ring

> **KEY POINT**
> Counter-urbanisation is the movement of people and economic activity out of larger settlements into the surrounding countryside and towns.

Land values are lower in these areas. Since the 1970s, housing, commerce and industry have moved out of the urban area to the settlements in the countryside. This is called counter-urbanisation. Quality of life is better in these areas; there is more space and less congestion, crime, vandalism and pollution.

Land use in settlements in LEDCs

The concentric zone and sector models were developed from studies of settlements in North America. They have been revised for cities in LEDCs.

- The function of the CBD remains the same as in the MEDC city. However, much of the area will be redeveloped with modern high-rise blocks, shopping malls, hotels and apartments. The area may contain one or two older buildings in an unusual style that are relics of a colonial past.
- The inner zone contains middle- and high-class housing. There will also be blocks of flats for the working class.
- The outer zone includes large houses and bungalows for the rich as well as squatter settlements.
- Industry is to be found along transport routes into or around the urban area.

Fig 9.8 Model of land use in an LEDC urban area

to coastal cities

Favelas: spontaneous shanty town, squatter-type settlements

Periferia: poor quality but permanent housing with some basic amenities

Modern factories along the main road favelas in between

Expensive, high rise flats with modern amenities

CBD: offices, shops and traffic congestion

Zone of average quality housing where better-off have moved out (19th-century)

Small, low cost, government house-improvement schemes

Modern factories along the main road

to coastal cities

High-class, suburban housing for executive and professional classes, with own commercial core

Factors affecting land use in settlements

Remember examples of these factors and mention that they can change, e.g. London Docklands became redundant as the size of ships increased but has been revived as a financial services and newspaper printing area.

What are the factors influencing the arrangements of **land use zones** in a settlement? An answer to this question allows planners to rearrange the land use in an attempt to improve the quality of life of the people living there.

The factors influencing land use in settlements include:

- accessibility
- competition for land
- the mixing of land uses
- the specialised needs of certain land uses.

Access

Different types of transport make some places in the settlement more accessible than others. For example, bus and trams routes allowed the countryside to become accessible because people did not have to walk to work anymore and the suburbs developed. Shops are located in very accessible places. This means the CBD, at ground floor level, is dominated by shops. However, as congestion has increased, larger shops have moved to the edges of settlements to retain their accessibility and provide space for car travellers to park.

Another example of **accessibility** was that housing used to be built next to factories. This still happens in some countries, e.g. Singapore. In the UK, industry has moved to more accessible locations on ring roads and motorways where they are close to settlements. People travel to these industrial estates in their cars.

Competition for land

Competition for land is at its highest in the CBD because of the good accessibility. Access is high because the centre is the focus of rail and bus routes. As a result, land prices rise. **High rents** mean only national chain stores, financial services, hotels and offices can afford the rents. Oxford Street and Regent Street in London and Fifth Avenue in New York are good examples.

A sign of the competition for each piece of land is the presence of **high-rise office blocks**. This means each piece of land can be used several times.

 KEY POINT

In some CBDs the competition for land has fallen as out-of-town centres have become more accessible and attractive. This has caused a decline of the CBD; shops have become empty or occupied by temporary bargain stores or charity shops.

Mixing of land uses

Some land uses do not mix. It is unlikely that high-income residential areas will be found close to heavy industry or airports. On the other hand, industries that need a large labour force, such as the clothing industry which requires large amounts of female labour, tend to locate near to large housing estates.

Pressure on land means that some new developments do mix unsuitable land uses. For example, new housing is build alongside new bypass roads but the houses have double-glazing and high fences to deflect the noise.

Specialised needs

Good examples of this include:

- shipyards on deep-water estuaries
- airports on extensive areas of flatter land
- warehouses and distribution depots near to motorway interchanges.

 PROGRESS CHECK

1. No settlements ever match an urban land use model perfectly. Why is it useful to study these urban models?
2. How can census data for small areas in a settlement be helpful to town planners?
3. Why is housing so expensive in the suburbs when land values are lower there?

1. Matching the model suggests a general explanation. Variations from the model suggest local factors. 2. Shows areas of deprivation and other groups with special needs. Also assesses housing quality. 3. Tends to be on larger plots of land with bigger houses. Demand is high, especially from families with wage earners.

LEARNING SUMMARY

After studying this section, you should be able to:

- **know the causes of urban growth**
- **understand the problems of urban areas**
- **recognise some planned solutions to urban problems**

Urban growth

 AQA A | AQA B
EDEXCEL B
OCR B | OCR C
WJEC A | WJEC B
NICCEA

In the last 150 years, urban areas have grown rapidly in population and area through **natural increase** and the **migration** of people from the countryside. Some recent examples are shown in Table 10.1.

Table 10.1

City	Population 1970 (millions)	Population 2000 (millions)
Mexico City	8.6	30.0
Sao Paulo	7.1	25.0
Calcutta	7.0	16.6

Fig 10.1 Projected urban population growth

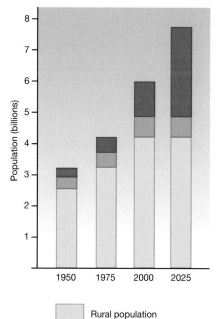

Rural population

Urban population

Developing countries

Developed countries

Mexico City occupied 50 km² in 1950 and grew to 1100 km² by 2000. Urban growth is taking place throughout the world. In 1950 the world's urban population was thought to be one billion but by 2000 it was estimated to be over four billion. Most of the increase today is taking place in LEDCs.

In 1900 only London and Paris had populations greater than one million (**millionaire cities**). By 2000 there were over 300 of these. Cities with more than eight million people were called **megacities**. In 1950 there were only two megacities: New York and London. By 1990 there were 21 megacities and 16 of them were in LEDCs. Recently the definition of a megacity has been changed to a city with a population of ten million, and in 1995 there were 15 megacities, with 11 of them in LEDCs. At current rates of growth there are expected to be 33 megacities by the year 2025, with 27 of them in LEDCs.

> **One billion is one thousand million (1 000 000 000).**

Fig 10.2 The world's megacities

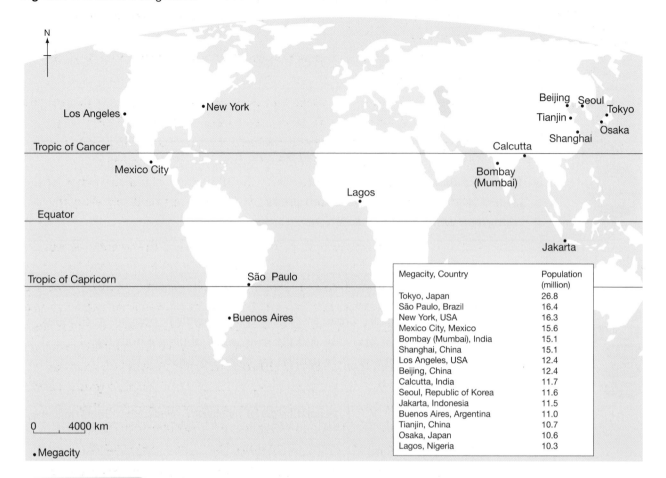

Megacity, Country	Population (million)
Tokyo, Japan	26.8
São Paulo, Brazil	16.4
New York, USA	16.3
Mexico City, Mexico	15.6
Bombay (Mumbai), India	15.1
Shanghai, China	15.1
Los Angeles, USA	12.4
Beijing, China	12.4
Calcutta, India	11.7
Seoul, Republic of Korea	11.6
Jakarta, Indonesia	11.5
Buenos Aires, Argentina	11.0
Tianjin, China	10.7
Osaka, Japan	10.6
Lagos, Nigeria	10.3

• Megacity

> Finding the total population of a city is difficult because it keeps changing. People come and go and its boundary moves.

Some countries, especially LEDCs, have one city that is much larger than all the others in the country. It is called a **primate city**. Such cities were usually established as the capital in colonial times. Primate cities include Lima (Peru) and Paris. The reasons for the development of primate cities include:

● high rural to urban migration
● the development of a core region containing the main economic activities
● the capital city of a centralised government
● when capital and major port are in the same city.

KEY POINT

The words conurbation or megalopolis are used to describe an urban area where a city has grown to absorb several larger towns, e.g. the Manchester conurbation.

The causes of urban growth

The three main causes of urbanisation are rural to urban migration, natural increase in the urban population and the redrawing of urban boundaries.

● Rural to urban migration (see page 110 in Chapter 8) occurred in MEDCs in the past as people moved to the new factories, ports and mines found in the urban area. Towns rapidly expanded to become cities. In LEDCs it is occurring today as people look for jobs, education and healthcare. Poor harvests, lack of agricultural land and farm mechanisation have pushed them away from the rural areas.

● Natural increase is taking place as healthcare becomes more widely available and medical advances combine with a better quality of life to increase life expectancy.

● Urban boundaries are redrawn as the urban areas expand to house all the people living there.

In the UK, cities spread as public transport improved. Suburbs of semi-detached houses developed on new housing estates. Light industry provided nearby employment or the new residents could **commute** to the CBD by bus and train. In the 1980s, business parks and out-of-town shopping centres encouraged this trend. However, since the 1950s, many local authorities have established green belts around the edge of the city to stop it sprawling any further across the countryside.

In the 1990s, better-off people, particularly families and the recently retired, moved out of the urban area to live in smaller settlements in the nearby countryside. They were moving away from the congestion, pollution, crime and noise of urban life. They found the rural environment more attractive and could travel quickly into the urban area if they needed to. This movement out of the urban area, which was first noticed in the USA in the 1970s, is called '**counter-urbanisation**'. Unfortunately this outward movement leaves behind lower income groups, including recent in-migrants from ethnic minority groups and the unemployed. Now some cities are attracting young people back by imaginative **redevelopment** of old factory and warehouse sites, e.g. Manchester and Leeds in the UK and Amsterdam in the Netherlands.

Whilst cities in MEDCs have tended to stop growing, those in LEDCs continue to grow and spread.

> **KEY POINT**
> Urbanisation is the process leading to an increased proportion of the people in urban areas.

Urbanisation

Urbanisation has been taking place for 200 years but there has been a dramatic increase in the last 50 years. It began in MEDCs with the Industrial Revolution; today between 70% and 80% of the world's population lives in urban areas (Table 10.2).

Table 10.2

Proportion of world's population living in urban areas	
1950	29%
1990	46%

In MEDCs the rate of urbanisation is slow because the majority of the population already live in urban areas. In addition, rural areas are more attractive and rates of natural increase are lower. Rates of urbanisation are much higher in LEDCs.

Urbanisation brings advantages and disadvantages. On the one hand more facilities such as education, healthcare and transport may develop. But disadvantages include many problems with inadequate services and lack of housing and jobs.

Problems caused by rapid urban growth

AQA A AQA B AQA C
EDEXCEL A EDEXCEL B
OCR A OCR B OCR C
WJEC A WJEC B
NICCEA

KEY POINT Rapid urban growth of population and city size causes a whole range of problems.

Housing

This is a common problem. Housing is often of poor quality, insufficient or too expensive. Where there is not enough housing, prices rise and the poor are forced to live in very low-quality houses. In some cities poor people build substandard shelters in **squatter settlements** on the edges of urban areas. (These are called *kampungs* in Jakarta, Indonesia.) They use plants, corrugated zinc sheets, and other scrap materials. Because of the shortage of land these camps are often built on flood plains where there is a danger of flooding, on steep hillsides where mudslides occur or close to railways and airports where noise makes life difficult.

These **squatter settlements** have many problems including:
- very poor-quality housing
- poor or non-existent water supply
- unreliable electricity often taken illegally from the mains
- little or no sanitation
- poor drainage for rain and household waste water
- little or no waste collection service and rubbish dumped on open spaces
- overcrowding and high risk of disease (e.g. tuberculosis)
- little employment because the factories are too far away.

The lack of affordable housing also causes homelessness. At its worst these people live and sleep in the streets. They cover themselves at night with cardboard and old blankets. On cold wet nights they sleep in office doorways and hope to get some heat as it escapes from inside the building.

Transport

Many urban areas have serious transport problems. Bangkok is said to have the worst **traffic congestion** in the world. When traffic congestion occurs, average speeds fall and results in frustration, increasing costs, people being late for work or appointments and falling productivity. This is mainly the product of:
- the growth of vehicle ownership in recent years, e.g. 15% annual increase in Indonesia and especially Jakarta
- poor road layouts especially in rapidly growing LEDCs
- poor public transport with buses and railways unable to keep up with the demand, e.g. trains in Tokyo, Japan and London and buses in Calcutta, India and Harare (Zimbabwe) are packed in the rush hours
- the rolling stock of public transport, e.g. buses and trains, is worn out and breaks down frequently
- breakdowns and accidents cause additional delays and congestion as the network reaches capacity.

The M25 motorway around London is sometimes described as the largest car park in the world!

Unemployment

Many of the people migrating to cities are male. They come to find a job and then ask their families to join them. Many do not have the skills for urban jobs and find themselves unemployed. They join the informal economy and run their own small business providing goods and services. In Beijing, China, they may run their delivery business on the back of a bicycle and their office would be the pavement. Cheap mobile telephones help.

Pollution

Urban areas suffer problems of **air, land, water and noise pollution**.

The causes of air pollution are:
- **industrial activity**
- **increased numbers of vehicles on roads**.

These put dust particles and higher levels of dangerous gases, such as carbon monoxide and sulphur dioxide, into the air. These gases can damage the health of people in the area especially the elderly, the asthmatic, those with bronchitis and eye problems. Air pollution also increases the dangers of fog and traffic accidents. Los Angeles and Mexico City have become famous for their high levels of air pollution.

Land pollution occurs when domestic and industrial waste is left uncollected. For example, domestic rubbish is usually collected once a week and its accumulation can cause pollution. In squatter settlements there may be no proper collection of domestic rubbish. Some industrial areas have no safe way of disposing of dangerous wastes and dump it on open spaces. Official and unofficial landfill sites bring the problem of flies and diseases such as diarrhoea and malaria.

Water pollution can be a major problem in urban areas. Rivers and canals can become polluted by domestic and industrial waste. In some LEDCs only 40% of houses are connected to the sewerage system. Even in some MEDCs much waste material is dumped through long pipes into the sea. This can threaten wildlife and the safe use of tourist beaches.

Noise pollution comes from industry, construction, and traffic including trains, motor vehicles and aircraft. Noise pollution adds to the stress of living in crowded urban areas.

> Make a two-column table and summarise these problems and their planned solutions.

Water supply

Some urban areas have inadequate water supplies. This is especially true when rainfall is low and the settlement relies on groundwater supplies. In addition, large quantities of water are lost through leaking pipes. The rapid growth of population and industry increases the demand. In richer areas modern domestic appliances and swimming pools also increase demand. It is estimated that water demand increases at three times the rate of population increase.

In cities in LEDCs many people do not have piped water to their houses. They may have it delivered in expensive water tenders by private water sellers or use a communal tap (one tap may serve 500 people and only work for a few hours each day). In the worst cases people use polluted wells and open pipes. For example, in some Indian cities clothes are washed in the street from storm water drainage pipes.

Water can also be of poor quality when water treatment plants cannot cope with demand. Poor water quality brings health threats. In the UK people are buying more bottled water, as they dislike the chemical taste of tap water.

Solutions to the problems of urban areas

AQA A   AQA B   AQA C
EDEXCEL A   EDEXCEL B
OCR A   OCR B   OCR C
WJEC A   WJEC B
NICCEA

KEY POINT Urban areas have many problems that local and national governments attempt to solve through town and country planning.

Housing

There are housing problems in both MEDCs and LEDCs. The solutions in MEDCs have similarities and differences from those in LEDCs. The main **solutions to the housing problems** in the urban areas of MEDCs are:

- use brownfield sites such as empty houses, former factories, office blocks, flats above shops or railway sidings. This land is often in poor condition and expensive to restore. However it has many advantages including the facilities in the town within walking distance, existing infrastructure and saving green fields
- the extension of existing towns by bolting on small suburban estates, often within the line of a newly built bypass
- building a small number of houses in several villages by infilling or small developments
- building new settlements such as new towns where all the infrastructure is new and houses can be built to the latest standards
- selecting one key village to be developed to the size of a small town with additional schools, bus routes and all the other services provided.

When a local authority plans to develop housing in an area it consults the people living there and asks for their suggestions. These schemes for housing estates in villages and rural areas usually attract opposition from local people. They are called NIMBYs by some people (Not In My Back Yard). The reasons given for opposition include:

- lack of employment locally
- increase in cars travelling on narrow village roads
- children travelling to school will cause congestion twice per day
- the small amount of public transport will not be able to cope
- the houses will spoil the views and newcomers will not fit into the community.

In LEDCs the solutions are different and include:

- discouraging rural to urban migration by improving conditions in rural areas by developing growth poles with jobs, housing, schools and healthcare. In Indonesia, rural people have to get a permit before they are allowed to move to an urban area
- improving living conditions in squatter settlements by building paved roads and pavements, laying drainage systems, improving water and electricity supply, building schools, clinics and community facilities. Unfortunately improvements attract even more people

● building subsidised public housing which may be low-rise, medium-rise walk up, e.g. Kuala Lumpur, Malaysia, and high-rise, e.g. Hong Kong and Singapore

Find a case study of a planned new housing development in your local area and one in an LEDC.

● encouraging self-help housing through co-operation between the government, local authorities and the people. Cheap land sites are provided with the basic infrastructure of roads, drainage, water pipes and sewerage systems. People then buy cheap materials to build their own houses under supervision e.g. South Africa, Zimbabwe and Indonesia.

Transport

There is a wide range of solutions to this set of problems:

● **building** new roads and widening existing ones with the building of flyovers, tunnels and bypasses, e.g. Birmingham, Leeds, Kuala Lumpur (Malaysia)

● **maintaining** roads and improving central crash barriers, street lighting and co-ordinated traffic lights

● **controlling** the increase in vehicle numbers, especially cars since each carries so few people, by increasing the costs to motorist through taxes on petrol, road taxes, tolls on busy roads (these actions are very unpopular with voters)

● **encouraging** the use of public transport that makes a more efficient use of the available road space. This can be done by subsidising fares or building new and faster transit systems, e.g. the MRT in Singapore, the MTR in Hong Kong, the new tube lines in London, tram systems in Manchester and Sheffield and BART (Bay Area Rapid Transit) in San Francisco

Some countries have introduced traffic lights that show the number of seconds to go to the next change of the lights.

● **discouraging** unnecessary travel by making it more expensive, e.g. high costs of parking, taxes to enter restricted zones (Singapore), and encouraging the use of faxes, telephones and electronic mail.

Employment

Tourism is one of the keys to increased employment. Tourists bring much needed foreign currency and demand many services. Many city authorities are allowing the informal sector to grow. Jobs include shoe shiners, street vendors, newspaper sellers, guides, food and drink sellers, taxi drivers (cycles/motor cycles), and makers of pottery and crafts, soaps and ornaments.

Pollution

The main types of solution that governments can use are:

● **fines** for breakers of environmental laws, e.g. if factories or cars emit more than the allowed level of gases or people litter the streets or discharge chemicals into water sources

● **public education** about the causes and consequences of pollution and encouraging people to be more thoughtful, e.g. limited hours for construction working called 'making pollution an individual matter'

● improving **waste disposal**, e.g. better sewerage pipes, twice-daily refuse collection in busy cities, provision of plastic sacks and 'wheelie bins'.

Water supply

The main solutions are:

- agreeing contracts for supply from areas with plenty of water
- building more reservoirs to store water
- investing in desalination plants to convert seawater into fresh water (but this is very costly)
- stopping water pollution at source
- building more water treatment plants
- installing piped water into older housing and replacing lead and steel pipes
- conserving water through education programmes
- saving water by its re-use, e.g. using clothes washing water to flush toilets
- improved industrial technology to recycle water.

In addition to these plans it helps to reduce the problems if stable governments are in power. They are able to reduce corruption and unrest which in turn attracts foreign investors who are attracted by the low wages and lack of trade union activity. Stable governments are also able to obtain foreign aid to repair the infrastructure and provide skills training for the local people. Some city governments have found that when they involve local people in decision-making their schemes are more successful.

Urban change in the UK: problems and planned solutions

AQA A   AQA B   AQA C
EDEXCEL A   EDEXCEL B
OCR A   OCR B   OCR C
WJEC A   WJEC B
NICCEA

KEY POINT In the UK there are very strong laws to provide solutions to urban problems.

The regeneration of cities

In the 1960s, the areas around the centres of cities in the UK had many problems. They had become run-down and overcrowded. Housing, built in the nineteenth century, was very poor and cramped with people sharing damp living areas and few had central heating. In these areas factories had closed due to poor access and the need for more space; the factories that were left were old-fashioned and unemployment was growing. The better-off and more skilled people moved to the suburbs, often following industrial employment. The people left in the inner city suffered from multiple deprivations.

The census of population offers measures of deprivation including overcrowding, single parent families, people living alone, unemployment, lack of basic housing facilities and no car.

The first round of plans to solve the problems, called comprehensive redevelopment, involved urban **redevelopment or renewal**. Slum clearance took place and many terraced houses were demolished. Clearance spread beyond the inner city to include the older suburbs. The people living in these areas were re-housed in new estates on the outskirts or in tower blocks in the inner city. Sheffield with the Hyde Park and Park Hill redevelopments became famous for this policy.

However, these new inner city estates became problem areas as the blocks were poorly built, communities were disrupted and crime grew. The tower blocks did not have any sense of community, they were noisy, insecure with dangerous dark corners in stairwells and the lifts and communal central heating frequently did not work.

Later schemes **rehabilitated** the older properties with damp-proofing, central heating, bathrooms, new doors and window frames, cleared spaces for parking and made open space into parks. Some of the tower blocks were demolished and low-rise blocks turned into houses with gardens.

Several governments have introduced a range of schemes to help the people still living in the **inner city**. These schemes include the views of local people and have included more provision for job creation and community activities.

A good case study is Hulme in Manchester where the blocks of the 'Crescents' have been replaced by low-rise family housing. People are now moving to live in Hulme.

Governments have noted that there are large areas of unused and derelict land in cities (**brownfield land**). Developers are encouraged to redevelop this land rather than use greenfield sites.

One famous scheme is the **London Docklands**. This government scheme regenerated the area back to effective use and included:
- reclaiming derelict land
- building new roads, the Docklands Light Railway and London City Airport
- new water, gas and electricity services
- building office blocks such as Canary Wharf
- creating many new jobs
- building new homes and refurbishing older ones to increase the resident population
- constructing new shopping facilities and a technology college
- planting over 100 000 trees.

The scheme has been criticised as failing to meet the needs of the local people. The homes provided were very expensive and the jobs were in high-tech industries and unsuited to local low-skilled people.

In the 1990s some run-down areas of cities have become fashionable places to live. The old houses have been bought cheaply and renovated by a wealthier middle class. This is called gentrification. The areas chosen usually have good access to office jobs in the CBD.

Green belts and new towns

Green belts

In the 1930s town planners tried to keep the countryside close to the people living in London by preventing further urban expansion. They established a green belt around the urban area and made planning permission within it very difficult to obtain. In the 1950s nearly all UK cities established green belts with the added idea of providing space for recreation.

New towns

New towns were built in the 1950s around London, Glasgow and Birmingham:
- to help prevent urban sprawl
- to act as growth poles to draw industry away from the congested cities
- to relieve overcrowding and create better housing within the urban area
- to provide housing and jobs for people willing to move from the city slums.

Over **30 new towns** have been built including Harlow, Stevenage and Cumbernauld. Some new towns in the UK have been successful and this planning idea has spread all over the world.

In some places the government has **expanded an existing town** instead of building a new one. This has been successful in Swindon and Peterborough. Critics of green belts say that, although they preserved a green area around the urban area, development of housing and factories 'leapfrogged' them and moved into the countryside beyond. Great pressure to be allowed to build on the green belt continues, especially in the south east of England.

Government planning in LEDCs

A number of LEDC governments have planned schemes to ease the problems they have in cities. These include:

- **Brasilia**, the new capital city of Brazil completed in the 1960s, to act as a growth pole to pull development away from the cities of Rio de Janeiro and São Paulo in the south east of the country. Whilst the city has had some successes, there is still heavy pressure on the older cities
- **new towns** have been built in Singapore and Egypt to divert growth from Singapore City and Cairo. They have been successful in re-housing people from the overcrowded areas in the main city
- **growth poles** have been established in rural areas to reduce the push factors driving people into the cities. This strategy has been used in Brazil and Zimbabwe.

PROGRESS CHECK

1. Describe the pattern of megacities.
2. Define suburbanisation and counter-urbanisation.
3. Describe two problems resulting from urban growth and explain proposed solutions.

1. See page 126 2. See page 127 3. See pages 128-132

11 Energy, resources and their management

After studying this section, you should be able to:

LEARNING SUMMARY

- define renewable and non-renewable resources
- know the main sources of energy and their alternatives
- understand how pollution and global warming are caused
- define sustainable development
- know the main principles of Agenda 21

What are resources?

AQA A
EDEXCEL B
OCR C
WJEC A
NICCEA

When people use something from the Earth it becomes a **resource**. Until people started using petroleum and uranium they were not resources. Just as materials become resources so the process is reversible as materials are replaced by better ones. Coal has been replaced in many cases by oil.

Resources, energy and fuel

AQA A
WJEC A
NICCEA

Energy is used to run machines and to provide heat and light. Energy is stored in fuels, such as oil and coal, or in moving water and wind. Coal and oil are also raw material resources for the chemical industry. Today the main form of energy is electricity that can be derived from all of these fuels.

World consumption is already growing at a rate equivalent to the UK's entire energy market each year. This adds the equivalent of an entire USA market, currently the world's largest energy consumer, every ten years. Yet today two billion people are still without access to modern energy. As populations and cities grow, access to traditional energy forms becomes difficult, and local environments may be degraded.

The growing demand for resources and energy

AQA A   AQA C
OCR A   OCR C
WJEC A
NICCEA

The demand for energy grows as population increases and economic development spreads to more countries. The demand for resources and energy is not evenly spread across the world. MEDCs have only 25% of the population of the world but they use up to 80% of the energy produced. Some MEDCs are trying to reduce their energy demand through more efficient motor vehicles and machines and better insulation. Most LEDCs get their energy from the efforts of people or animals or through imported fuels because they do not have significant **reserves of energy** of their own. Demand in LEDCs is set to rise as countries industrialise and more people own cars.

Renewable and non-renewable resources and energy

Renewable resources include hydro-electric, solar, wind, wave, geothermal and tidal power. They do not become exhausted.

Non-renewable resources include coal, oil and gas. They were formed millions of years ago and are known as fossil fuels. They can become exhausted. Currently about 95% of the world's energy supply comes from non-renewable sources.

Some resources can be thought of as both renewable and non-renewable. Wood can be used for fuel and the trees replanted. In practice in LEDCs the trees are cut for fuel and not replanted so they become non-renewable. Nuclear power is also non-renewable but current levels of reserves will last at least 1000 years.

> Make a list of renewable and non-renewable resources. Be clear in your mind which resources can fit in either group.

> **KEY POINT** Energy resources are vital to the development of a country.

Energy, resources and their management

AQA B AQA C
EDEXCEL B
OCR A
WJEC A
NICCEA

Coal

Coal is a sedimentary rock and was formed by the compression of trees and plants many millions of years ago. In the past it was the main fuel for industry, especially for iron and steel making, transport, particularly steam railways, and for the heating of homes. Today it is mainly used in power stations to produce electricity.

> Large deposits of coal are found in Yorkshire, Derbyshire and Nottinghamshire, north-eastern France and Germany, the north-east USA, New South Wales in Australia, South Africa and India.

The main features of the **coal mining** industry are:
- it can be mined underground (shaft mining) or on the surface (open-cast)
- there are still 300 years of reserves for this relatively efficient fuel
- coal mines create visual, noise and air pollution
- mining can be dangerous and accidents do occur
- coal is bulky and costly to transport
- burning coal creates acid rain and adds to global warming.

Oil and natural gas

Fig 11.1 An oil field

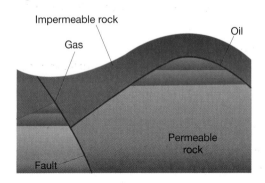

Oil and natural gas are formed from the remains of tiny sea creatures that died and were buried millions of years ago. Oil and natural gas are found in pockets beneath a layer of impermeable rock. The major oil fields are found in the Middle East, e.g. Saudi Arabia and Kuwait, Alaska and under the North Sea.

The main features of the **oil and gas industry** are:
- both fuels can be burned in power stations to produce electricity
- the fuels are extracted by drilling wells
- drilling can be dangerous, e.g. explosions and fire, in difficult environments such as Alaska and the North Sea

Note how OPEC controls the supply of oil and the effect this has on the economies of MEDCs.

- drilling in the North Sea takes place from massive oil rigs
- oil and natural gas are easier to transport than coal, using pipelines and massive sea-going tankers
- oil produces many valuable raw materials for other industries
- burning these fuels produces less greenhouse gases than coal
- known reserves of oil and natural gas will be used up by 2050
- serious environmental damage is caused if a pipeline breaks or a tanker is grounded, e.g. the *Exxon Valdez* in 1989
- developing an oil or natural gas resource creates wealth in the area
- oil has become a political weapon with the oil-producing and oil-exporting countries (OPEC) limiting supplies to maintain higher prices
- large reserves of oil are found in politically unstable areas, e.g. Iraq.

Nuclear energy

Nuclear energy is stored in uranium atoms and is released as heat to turn water into steam that drives steam turbines creating electricity. The energy in a nuclear power station is created when neutrons strike uranium 235 in a process called nuclear fission. Nuclear energy was seen as the source of energy for the future. **Accidents**, such as those at Five-Mile Island, Pennsylvania in 1979 and at Chernobyl in Russia in 1986, have caused many to question its use in the future. In the UK, nuclear stations were built on the coast at Hinckley in Devon, Sizewell in Suffolk, Hunterston near Glasgow, Dungeness in Kent, Dounreay in the north of Scotland and Wylfa on Anglesey. One problem for nuclear power stations is decommissioning at the end of their lives.

France has few fossil fuel resources and despite its hydro-electricity capacity has turned to nuclear power. Globally, there are about 500 nuclear reactors producing nearly 20% of the world's energy supply.

List the advantages and disadvantages of using nuclear power.

Views are divided on the future for nuclear power. The technology has the ability to create almost unlimited cheap power with relatively low environmental risks. Those against nuclear power point to the accidents, the dangers of military use, its inflexibility to meet peak demand and problems of nuclear waste disposal.

The main characteristics of the **nuclear energy industry** are:
- uranium reserves will last for many years
- refined uranium is relatively easy to transport
- power station sites do not need to be located near to the raw material source
- nuclear stations need large amounts of water and so tend to be located near to coasts or estuaries
- nuclear power stations produce no greenhouse gases
- it is easier for countries with low fossil fuel reserves to import small amounts of uranium to set up power stations
- there is widespread public concern about the safety of nuclear power both in terms of the transport of fuel as well as local radiation effects (unexplained clusters of cancer cases have occurred near some major nuclear stations)
- nuclear waste remains radioactive and dangerous for many hundreds of years.

Alternative sources of energy

AQA A AQA B AQA C
EDEXCEL B
OCR A
WJEC A
NICCEA

Using fossil fuels to generate energy does harm to the environment and these fuels will run out in the future. Nuclear energy is a politically unpopular source of power. For these reasons governments and non-governmental organisations (NGOs) are developing alternative sources of power. They aim to find and harness efficient, cleaner and more permanent sources of power. The main alternative sources of energy are hydro-electricity, the wind, the sun (solar power), the waves and the tides.

Hydro-electric power (HEP)

Fig 11.2 Cross-section through an HEP station

The power for **HEP** comes from moving water as it passes over a waterfall (Niagara Falls) or is released from behind a dam (Kariba Dam). The water drives turbines that generate electricity. Steep-sided valleys or gorges cut in strong impermeable rock in areas of reliable rainfall provide good sites. The huge scheme to dam the Yangtse in its gorge section has been criticised for the environmental damage it will cause with so many people losing their homes.

The main characteristics of **hydro-electric power** are:
- it does not produce any greenhouse gases
- once the dam is constructed, it provides relatively cheap electricity
- the dam controls the flow of water and can be used to prevent flooding downstream
- lakes behind the dam can be used for recreation and wildlife sanctuaries
- the dam is expensive to build and often difficult to access
- a large area of land may be flooded and people displaced (e.g. the Volta Dam in Ghana)
- the reservoir lake behind the dam can slowly silt up.

Fig 11.3 A wind turbine

Wind power

On hillsides in the north of England large numbers of three-winged windmills are being used to generate electricity. The wind blows onto the rotor blades and as they rotate they generate electricity. Many parts of the UK are suitable for '**wind farms**' because they experience reliable winds.

The main characteristics of wind power are:
- the electricity produced is cheap and does not cause pollution
- sites can either take a small number of turbines, to serve a few houses, or several hundred
- windy sites are unfortunately often in attractive upland areas
- turbines are noisy for people living nearby and a large number would be needed to make any major impact on the demand for fossil fuels.

Solar power

To produce solar power, a panel containing light-sensitive cells is exposed to **sunlight**. The cells can be used to heat water or to convert the sunlight to electricity. The panels are placed on the roofs of houses and in sunny, cloudless areas provide power to the house.

The main features of solar power are:
- it is mobile and is used to power electronic equipment
- satellites and other space stations use solar power to extend their effective lives
- its use is limited by night and by clouds
- the technology is only 15% efficient at converting sunlight into energy.

Energy and the environment: a case study from Alaska

Oil in Alaska

In 1968 the largest oilfield in North America was discovered in Prudhoe Bay, Alaska. The field is north of the Arctic Circle in an area where the climate is harsh. The field is rich in oil and gas but it is remote from users in the United States. It would have been difficult to use oil tankers to move the oil because the seas are dangerous and frozen for parts of the year. The alternative was to build a pipeline, 1250 km long, across Alaska to the ice-free port of Valdez. This would mean building a pipeline across seven ranges of mountains where the ground is frozen for much of the year. The pipeline would cross one of the last true wilderness areas in the USA. There was opposition from environmental groups but by 1977 the pipeline was built.

The problems faced by the pipeline builders included:
- the area was isolated with no road access until 1968
- low temperatures in winter make working conditions dangerous and the oil flow slowly
- up to two metres of snowfall in winter
- fogs and gales are frequent on the coasts
- the ground is frozen so that the pipeline could not go underground
- in the summer the surface soil thaws and may flow causing damage to the pipeline
- buildings tend to sink into the soil as they warm and melt the frozen soil
- the vegetation would be destroyed during construction
- caribou migration routes could be blocked
- oil leaks on land and sea could damage unique environments
- the wilderness would be lost
- local people might be forced to move.

These problems were reduced by:
- building the road on gravel pads to increase drainage and reduce the dangers of ice
- putting the pipeline on three-metre-high stilts, with the pipe resting on sliding shoes to reduce the danger of melting the soil and reduce damage through seasonal movement
- lifting the pipeline to six metres where it crossed migration routes

● building twelve monitoring stations along the length to check for leaks.

Since oil began flowing there have been over 300 reported small leaks and a major disaster when the *Exxon Valdez* oil tanker ran aground. Its oil cargo caused massive environmental damage.

The Prudhoe Bay oil field is beginning to run dry and the oil companies want to extend their explorations for more reserves. The arguments for and against are explored in Table 11.1.

Table 11.1

For oil exploration in Alaska	Against further exploration
Brings jobs and income to Alaskans	Environmental groups regard the fragile
Money can be used to build schools	environment of Alaska as unique
and hospitals in Alaska	Some Alaskans prefer the area as it
Roads and water supply can be	was before oil was extracted
improved	There are dangers of another serious
Provides oil and gas for US industry	oil spillage
Interest in the Alaskan environment is	Some believe that oil companies do not
only a minority interest	protect the environment
Reduces the dependency of the US on	Finding more oil delays the time when
oil from foreign countries who may be	real action will have to be taken about
hostile	global warming

Similar case studies could be found concerning the quarrying of limestone for cement in Derbyshire or the felling of trees for hardwood in the tropical rainforests.

Pollution

Pollution topics to study:
● burning fossil fuels
● destruction of the rainforest
● freshwater and marine water pollution
● acid rain
● causes, consequences and strategies to alleviate pollution including international co-operation

Possible causes of and solutions to global warming

AQA A   AQA B   AQA C
EDEXCEL B
OCR A   OCR C
WJEC A
NICCEA

Global warming: the evidence

Rising temperatures and warmer weather since 1980 suggest that global warming is taking place. Some scientists predict a rise of 3°C in the next 100 years. Possible effects of global warming include:
● rising sea levels as the oceans expand at higher temperatures and the ice caps melt. Areas at risk would include the east coast of England, Bangladesh, the Nile delta and the Netherlands
● greater evaporation and increased rainfall
● increased plant growth where higher temperatures and water are available

> Do not confuse global warming and greenhouse gases with the increased threat of skin cancer because of the increasing hole in the ozone layer from CFCs.

- diseases such as malaria, yellow fever and dengue fever spread further north and south
- more changeable weather with stronger storms threatening coastal cities
- low-lying areas such as Florida, the Netherlands and Lincolnshire under threat from rising tides and storms
- some cereal-growing areas and some poorer countries in Africa becoming drier
- loss of many birds and other animals including the polar bear as their habitat disappears.

No one can be certain that global warming is taking place. The higher temperatures could be another natural climatic variation, a fluctuation in solar output or the product of dust from volcanic eruptions.

The greenhouse effect

During the day the Earth is warmed by in-coming short-wave radiation from the sun (surface temperature of 5000°C). At night the Earth cools as it emits longer-wave radiation. Not all of the heat reaches space. Some is reflected back by clouds and some gases in the atmosphere which trap heat in the same way that a greenhouse would in a garden. Hence they are called '**greenhouse gases**'.

These greenhouse gases occur naturally and keep the temperature of the Earth 33°C higher. (Note during the Ice Age the temperature only fell by 4°C.) The amount of greenhouse gases in the atmosphere has been increasing as people burn fossil fuels, burn and clear rainforests, increase landfill sites, drive more miles, use refrigerators and aerosol sprays. This increase in greenhouse gases is trapping more of the outgoing long-wave radiation and reflecting it back to the Earth. More and more heat becomes trapped in the atmosphere of the Earth.

Sustainable development

AQA A · AQA B · AQA C · EDEXCEL B · OCR A · OCR B · OCR C · WJEC A · WJEC C · NICCEA

KEY POINT Sustainable development meets the needs of the present generation and retains our ability to meet the needs of future generations.

Sustainable development can be promoted through conservation, resource substitution, recycling, use of appropriate technology, pollution control, using renewable energy sources such as wind and solar power, international action by groups of countries, and lower consumption of energy through better insulation, using fluorescent light tubes, using smaller cars, and cycling and using public transport instead of making car journeys.

Some planners in the UK have added to the definition of sustainable development. They include actions to ensure a settlement survives and is sustained. For example, they want to build affordable houses for young families in a village where the school is threatened with closure.

Agenda 21

In 1992 the **United Nations** produced Agenda 21 on sustainability with the slogan 'think globally and act locally'. The agenda has seven key points:

- interdependence – people, the economy and the environment depend upon each other locally and globally
- citizenship and stewardship – people have rights and responsibilities including participation and co-operation
- regard for the future – responsibility includes protecting future generations
- diversity – recognition and respect for the diversity of cultures, societies and environments on Earth
- equity – respect for people's quality of life with justice and fairness for all
- sustainable change – planning and monitoring change to ensure the future
- uncertainty – understanding the uncertainty of the future and taking appropriate precautions.

Examples of **sustainable development** are:

- the use of brownfield land rather than green fields for new housing, factories and business parks
- the farming of forests so that more trees are planted than are being removed
- the imposition of quotas on sea fishing
- the provision of cycle tracks in urban areas
- improvements in electrically-driven public transport
- park-and-ride schemes (this is disputed by some who feel they encourage car use)
- organic farming using manure rather than chemical fertilisers
- the use of alternative energy sources instead of fossil fuels.

PROGRESS CHECK

1. What is the difference between renewable and non-renewable resources?
2. Suggest reasons for the lack of popularity for the development of nuclear power.
3. Describe and explain what is meant by sustainable development.

1. See page 136 2. See page 137 3. See page 141

After studying this section, you should be able to:

- *understand farming as a system*
- *locate the different types of farming in the UK*
- *understand the factors affecting farming*
- *describe different types of farming*
- *recognise the ways in which farming is changing*

Farming, fishing, mining and forestry are called primary industries. They involve extracting resources from the Earth.

Farming as a system

AQA A   **AQA B**
EDEXCEL A   **EDEXCEL B**
OCR C
WJEC A
NICCEA

Each type of farming can be described as a system with inputs, processes and outputs (Fig 12.1).

Fig 12.1 Systems diagram of agricultural activity

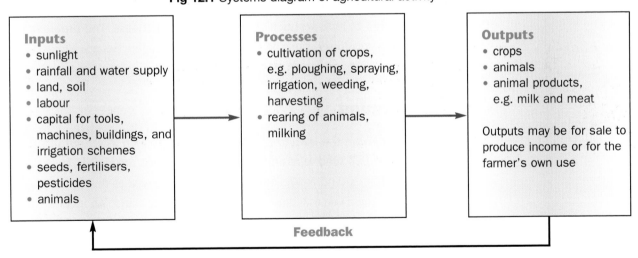

Inputs
- sunlight
- rainfall and water supply
- land, soil
- labour
- capital for tools, machines, buildings, and irrigation schemes
- seeds, fertilisers, pesticides
- animals

Processes
- cultivation of crops, e.g. ploughing, spraying, irrigation, weeding, harvesting
- rearing of animals, milking

Outputs
- crops
- animals
- animal products, e.g. milk and meat

Outputs may be for sale to produce income or for the farmer's own use

Feedback

Feedback from the system

- Outputs produce profit for use on the farm
- Experience determines next year's decisions
- News from the government (NFU and MAFF) and media suggests changes
- Some output is retained as animal feed

Distribution of farming types in the UK: the four main types

KEY POINT Each part of the UK tends to have a particular type of farming.

Fig 12.2 The main farming types in the UK

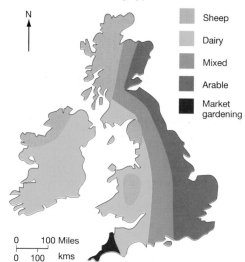

Key:
- Sheep
- Dairy
- Mixed
- Arable
- Market gardening

0   100 Miles
0   100   kms

> It will be helpful if you know five case studies, one for each of these types of farming.

Dairy farming

This involves the rearing or purchase of cattle for the production of milk. It is mainly found in south-western England, the lowland areas of Wales and in Lancashire. Dairy farmers prefer fertile, well-drained soils that produce high quality grass. Dairy farms supply milk to nearby urban areas as well as to dairies for the production of milk products such as butter and cheese. The production of milk has been subsidised since 1945 but the quantity farmers are allowed to produce has been limited by **quotas** since the 1980s.

Arable farming

Arable farmers plough the land to produce cereal crops such as wheat and barley, vegetables, oil seed rape and linseed. Arable farms are mainly found in eastern England, including Norfolk and Lincolnshire, as well as the east of Scotland. The farmers use machinery and so prefer flatter land. An ideal climate has warm summers with rain during the summer growing season. Arable crops attract guaranteed prices through the **Common Agricultural Policy** (CAP) of the European Union.

Sheep farming

Sheep produce meat and wool. Sheep farms, which are mainly family run, are found in the upland marginal and peripheral areas of England, Wales and Scotland. Sheep farming can take place on steep slopes, with thin soils, high rainfall and low temperatures. The sheep feed on the poor grass that grows on the moorlands. Hill sheep farms receive subsidies from the EU and some farmers earn extra income from campers and caravan users. The lambs produced each February and March are taken to lowland farms to be fattened for sale as meat.

Market gardening

This is **intensive** farming, producing high-quality crops such as fruit, salad, vegetables and flowers. It can be found in Cornwall and Devon, the Isles of Scilly, and the Fens. Farms may be as small as one hectare. Market gardeners use climate-controlled glasshouses, artificial soils and irrigation to obtain a high quality output which is taken to markets in refrigerated lorries.

Factors affecting farming

Farmers make decisions about what to grow, what animals to keep, the level and type of inputs and the methods they will use. Their decisions are based upon a range of social, economic and environmental factors. The farmers' attitudes and level of knowledge are also important.

Social and economic factors

The importance of each factor varies from farm to farm. Note how each influences farms near to you.

These are human factors and include labour, capital, technology, markets and government (political).

Labour
- In LEDCs, such as India and Java, farmers use abundant cheap labour instead of machines. In Japan and the UK, where labour is expensive, they use machines.
- People working on farms may be unskilled labourers or skilled and able to use machinery, e.g. tractors, harvesters and milking machines.

Capital (finance)
- Capital, the money the farmer has to invest in the farm, can be used to increase the amount of inputs into the farm, e.g. machinery, fences, seeds, fertiliser and renewing buildings.
- If a farmer can afford to invest capital, yields will rise and can create greater profits which can be used for more investment.

Technology
- Machines and irrigation are two types of technology that can increase yields.
- Greenhouses, with computer-controlled technology, provide ideal conditions for high quality crops. The computer controls the temperature, moisture level and amount of feed for the plants.
- Genetic engineering has allowed new plants to be bred that resist drought and disease and give higher yields.

Markets
- Farmers grow crops which are in demand and change to meet new demands, e.g. rubber plantation farmers in Malaysia have switched to oil palm as the demand for rubber has fallen.
- Markets vary throughout the year and farmers change their production to suit them.

Government
- Governments influence the crops farmers grow through regulations, subsidies and quotas.

> Think of the farmer weighing up all these factors when deciding what to plant. Some richer farmers have plenty of choice.

- Governments offer advice, training and finance to farmers and, in new farming areas, may build the infrastructure of roads and drainage, e.g. Amazonia.
- In some countries, e.g. Kenya and Malaysia, the government is trying to help nomadic farmers to settle in one place.
- Some governments plan and fund land reclamation and improvement schemes.

Environmental factors

These are **physical factors** and include climate, relief and soil.

Climate

- Temperature (minimum 6°C for crops to grow) and rainfall (at least 250 to 500mm) influence the types of crops that can be grown, e.g. hot, wet tropical areas favour rice, while cooler, drier areas favour wheat.
- The length of the **growing season** also influences the crops grown, e.g. wheat needs 90 days. Some rice-growing areas have two or three crops per year.

Relief

- Lowlands, such as flood plains, are good for crops.
- Steep slopes hinder machinery and have thinner soils; lower, more gentle slopes are less prone to soil erosion.
- Tea and coffee crops prefer the well-drained soil on hill slopes.
- Temperature decreases by 6.5°C for every 1000 metres gained in height.
- South-facing slopes receive more sunlight.

Soil

- **Fertility** is important; poor soil means lower outputs or larger inputs of fertilisers.
- Floodplains are good for rice because of the alluvial soils.
- Good drainage reduces the dangers of waterlogging.

PROGRESS CHECK

1. Name the five main types of farming in the UK and for each identify two main characteristics.
2. Describe the difference between social and economic factors and environmental factors influencing farmers' decisions.
3. Describe how governments can affect farming in LEDCs and MEDCs.

1. See page 144 2. See pages 145–6 3. See page 145–6

Different types of agricultural activity

AQA A   AQA B
EDEXCEL A
OCR A
NICCEA

There are many different types of agriculture including:

- shifting cultivation, e.g. the Amazon Basin in Brazil, Sabah and Sarawak in Malaysia
- wet rice cultivation, e.g. Ganges Plain in India, Chang Jiang Plain in China, southern Japan, Java in Indonesia

Use an atlas to locate the many places named in this section.

- market gardening, e.g. Vale of Evesham in the UK, the Netherlands, around Singapore
- plantation agriculture, e.g. west coast Malaysia for oil palm and tea.

> **KEY POINT** Subsistence farmers grow crops and rear animals mainly for their own use.

Shifting cultivation

Shifting cultivation is under threat as population grows and land becomes more scarce.

Shifting cultivation occurs in South-East Asia, Central and South America and Africa. The cultivators cultivate more remote area such as equatorial forests (Amazon) and highlands (Sarawak). They grow crops for themselves and their families. The main crops are rice, maize, tapioca, sweet potatoes, bananas and vegetables.

The main characteristics of shifting cultivators are:

- low farm input with manual labour and simple tools
- low yields, and when the soil loses its fertility, the land is abandoned and may suffer soil erosion
- low technology such as axes and knives to clear forest; the felled trees are burned although shifting cultivators in Sarawak now have chain saws
- low output, mainly food for the family
- many crops are grown, up to 30 in a small clearing, to provide food throughout the year
- governments are attempting to persuade shifting cultivators to settle on farms.

Fig 12.3 Distribution of shifting cultivation

N

Tropic of Cancer

Equator

Tropic of Capricorn

0   4000 km

Shifting cultivation

Wet rice farming

One-third of the world's population depends on rice for food. Many Asian countries grow rice. In India and China it is grown as a subsistence crop whilst in Japan and Thailand it is grown commercially. Climate, relief and soil influence the distribution of rice growing areas. Wet rice requires 1000mm to 2500mm of rainfall a year and an average temperature of 20°C. Wet rice is grown on the fertile alluvial soils of the Ganges and Chang Jiang Plains. Where flat land is limited, farmers terrace hillsides e.g. Java (Indonesia) and Luzon (Philippines). The bund (raised bank of soil) at the edge of each terrace retains the rainwater.

The main characteristics of wet rice cultivation are:

- high inputs of fertilisers and labour
- irrigation allows two or even three crops per year (e.g. Philippines and Vietnam)
- the level of technology varies with each area; low in India and China with mainly manual labour and water buffaloes but high in Japan and Taiwan with modern irrigation and machines
- small farms (less than two hectares) are fragmented through inheritance laws which demand that the land is divided between the sons
- rice is the main crop though some vegetables are grown
- planting and transplanting are usually done by hand.

Fig 12.4 Wet rice cultivation in Asia

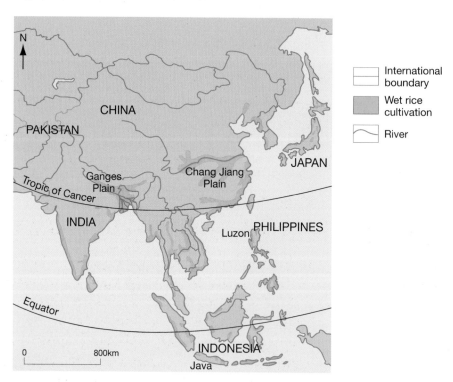

Commercial farming

Crops that are grown for sale are called cash crops, e.g. wheat and barley for food, cotton and rubber as raw materials for industry. Animals, such as sheep and cattle, are also bred for sale.

> **KEY POINT**
>
> Commercial farmers grow crops and rear animals for sale to make a profit. Intensive farming uses high inputs, such as labour and capital, per unit area.

Intensive commercial farming: market gardening

Intensive farms have high **inputs** per hectare (area of land), e.g. intensive market gardening in the Netherlands. These farms produce the vegetables, fruit and flowers that are found in supermarkets. Market gardening is often found on the outskirts of urban areas. In the past, people living in urban areas produced very little food for themselves and market gardening crops used to perish rapidly before advances in transport and refrigeration.

A case study of a market garden is often useful in farming and urban questions.

The main characteristics of **market gardening** are:

- perishable farm outputs are sold to nearby urban areas but refrigeration and faster transport allow more distant markets to be served
- affluent people demand products grown all over the world all the year around
- market gardens have high inputs, especially labour, high yields and profits
- market gardens use limited land and those close to urban areas are under threat from building developments
- a wide range of technology is available from hand hoes to automatic water sprinklers and computer-controlled robots
- the level of output is very varied from low-output peasant farms to high-output company farms
- farmers usually specialise in a few crops, e.g. salads or flowers
- many market gardens are moving to organic production (minimum use of chemicals) to meet growing demand from urban areas
- Singapore has established 'agrotechnology' parks for intensive specialist vegetable, poultry and fish farming
- some salad vegetables are grown without soil using hydroponics (plants grow on a perforated plastic sheet and are fed liquid foods).

Extensive commercial farming

MEDC agriculture case study: the Prairies of North America

Use your atlas to find this area of farming.

The western coast of North America is fringed by the Rocky Mountains. To the east of these mountains there are extensive grasslands. They stretch from Texas in the south to Alberta and Saskatchewan in the north and are called the Prairies. This is a rich area of extensive farming. The eastern parts of these vast plains are famous for their production of cereals because:

- precipitation is over 1000mm which allows good grain production
- dark chernozem soils are rich in humus and are well drained
- spring and winter wheat and maize grow well
- farms are large and highly mechanised
- railways were built to transport grain to eastern ports in Canada
- the extensive rolling landscape with large fields is ideal for mechanisation.

Fig 12.5 Extensive commercial farming in the Prairies

Summer temperatures average 19-21°C; Precipitation 250-500mm; heavy winter snow; strong winds cause loss of moisture; light brown soils

Precipitation over 1000mm; dark soils with high humus content; chernozems in best areas. Diverse native vegetation, much now cleared for cereals (spring wheat in north; winter-sown wheat in centre; maize in south where growing season is longer)

Summer temperatures 28°C; precipitation over 500mm; summer convection rainfall; heavy winter snow; chestnut soils extensively cultivated in North and South Dakota, Nebraska and Kansas

Key
- Present-day extent of open prairie
- International boundary
- Boundary of states

Agriculture on the Prairies is changing in the following ways:

- farmers are growing crops in rotations, which include peas and barley to protect the soil
- global warming is threatening to raise temperatures and reduce rainfall
- long stubble fallow is used to conserve moisture in the soil
- farmers are diversifying into non-agricultural activities including leisure and tourism.

Plantation agriculture

> Plantations that grow trees for timber or wood pulp are not usually accepted as 'plantation agriculture'.

Plantation agriculture takes place on large farms or estates. These estates were developed in South-East Asia and the Caribbean during the eighteenth and nineteenth century colonial period. During this time European countries, such as the UK and the Netherlands, set up plantations to grow rubber, coffee, tea, sugar and spices. The hot wet climate favoured these crops which were exported to meet demand in Europe.

The main characteristics of plantation agriculture are:
- crops are grown commercially for export or for use in local industries

- plantations need very large amounts of capital for road building, planting and the establishment of processing factories
- large numbers of workers are employed in administration, technical services and labouring
- high levels of technology are used, e.g. machines for planting and harvesting as well as in laboratories to check the quality of the product
- plantations are large (40 hectares to 1000 hectares) and produce a massive output but usually of only one crop
- tree crops are usual, e.g. tea, rubber and oil palm; they produce crops throughout the year.

Agricultural change

AQA A AQA B
EDEXCEL A EDEXCEL B
OCR A OCR B OCR C
WJEC B
NICCEA

KEY POINT The Green Revolution involves the use of high-yielding varieties of seed, increased irrigation and the use of fertilisers to improve food supply.

Meeting the rising demand for food: the Green Revolution

During the 1960s, as world population grew, there was increasing concern in many countries about providing an adequate food supply. The main plan to increase food supply was called the Green Revolution.

Countries such as India and Indonesia made great efforts to increase the supply of rice and wheat using irrigation, new seeds and fertilisers. The main aspects of the Green Revolution are:

> The Green Revolution can be included in questions on agriculture or food supply. It is very popular with examiners.

- use of high-yielding varieties (HYVs) of seed produced by scientists in the Philippines. The first successful HYV was IR-8 the 'miracle rice' that doubled yields
- larger amounts of fertiliser are needed by the HYVs
- HYVs need adequate and carefully-controlled water supplies
- further development of HYVs has reduced the growing period from 180 to 100 days
- scientists are continuing to develop new varieties of seed that need less irrigation, are resistant to disease and are good to eat
- governments are offering loans, advice, storage and transport facilities to poorer farmers.

The Green Revolution has brought benefits and problems.

Benefits
- Increased rice production; the total produced doubled, especially in China.
- Improved standard of living as farmers sold their surplus, e.g. the Punjab in India where farmers were willing to accept the changes.

Problems
- Irrigation is essential for the best results from HYVs.
- HYVs are not successful on alluvial plains where flooding occurs.
- HYVs are more costly to grow, needing more fertiliser and irrigation.
- Farmers who can afford to grow HYVs get richer and this increases the gap between them and the poorer farmers.

Use a table to list advantages and disadvantages.

- Fertilisers for HYVs are creating pollution in rivers and lakes.
- HYVs have been vulnerable to outbreaks of disease.
- Mechanisation has increased unemployment.

The European Union (EU) and its Common Agricultural Policy (CAP)

The **CAP** has been in action, for member states including the UK, since 1962. It aims are:

- to protect the income of farmers
- to ensure reasonable and steady prices for consumers
- to ensure food supplies through increased production
- to protect the quality of life in rural areas.

These policies have been put into action through grants, subsidies and guaranteed prices. Stable prices allow farmers to plan. The policy has been criticised for protecting weaker farmers, creating mountains of food and lakes of wine, making food expensive when compared with the USA and for environmental damage (hedgerow removal, soil erosion, excessive use of fertilisers, pesticides and herbicides) as farmers attempt to increase production. Recently the EU has been paying farmers to take part of their land out of production (set-aside).

The CAP is unpopular and seen as protecting poorly-organised farmers at the expense of the public. It protects European farmers against cheaper food from LEDCs.

Farm diversification

The National Farmers Union has produced a list of nearly 30 ways in which farmers can use their land to increase their income.

Farmers in the UK have found it increasingly difficult to make their farms economic. Food surpluses, milk quotas, reduced subsidies and growing public awareness of environmental issues have combined to reduce farm income. Farmers have been encouraged by government to move to other means of earning income (diversification).

Farmers have turned to:

- pick-your-own (PYO) for apples, strawberries, gooseberries and other fruit
- garden centres and farm shops for the sale of farm produce and garden plants
- barns and outbuildings converted to holiday cottages
- small industrial estates near to the farm buildings
- golf courses and nature reserves open to the public on farm land.

PROGRESS CHECK

1. Describe the main inputs to a farm.
2. What are the five main types of agriculture found in the UK?
3. List the human and physical factors influencing a farmer's decisions.

1. See page 143 2. See page 144 3. See pages 145–6

LEARNING SUMMARY

After studying this section you should be able to:

● **define primary, secondary and tertiary industry**
● **understand industry as a system**
● **know the main factors of industrial location**
● **describe industry in LEDCs and MEDCs**
● **understand the rise of the Newly Industrialised Countries (NICs)**
● **understand the changing structure of industry**
● **recognise change in the location of shops**

Different types of industry

Industry can be divided into three main types: primary, secondary and tertiary industry. Recently, tertiary industry has been sub-divided to give a fourth type: quaternary industry. Sometimes secondary industry is called manufacturing industry and tertiary is called services.

> **KEY POINT**
>
> An occupation is the job that someone does; employment is the industry in which they work. A person can have the occupation of accountant and be employed in a car assembly plant.

> A triangular graph is a useful way of showing the balance of primary, secondary and tertiary industry in a country.

Primary industry

Primary industry involves extracting resources from the sea or land and includes farming, fishing, forestry, coal mining, oil drilling and hunting. It is located where the raw material is available. Primary industries are particularly dependent on physical factors such as climate and geology as well as government intervention, e.g. the Common Agricultural Policy of the EU.

Secondary industry

Secondary or manufacturing industries make things for people by processing raw materials or assembling components. The raw materials may be obtained from primary industry or be the products of other secondary industries. A tin of fruit or a motor car are manufactured by a secondary industry using the products of primary and manufacturing industries.

Tertiary industry

Tertiary industries provide a service and include doctors, teachers, shop assistants, entertainers and lawyers. Health, administration, retailing and transport are called service industries.

Quaternary industry

Quaternary industry is concerned with research and development as well as information and communications technology which help companies to function.

Industry as a system

 AQA A   AQA B
OCR A   OCR C
NICCEA

Manufacturing industry can be seen as a system with inputs, processes and outputs (Fig 13.1).

Fig 13.1 Systems diagram of manufacturing industry

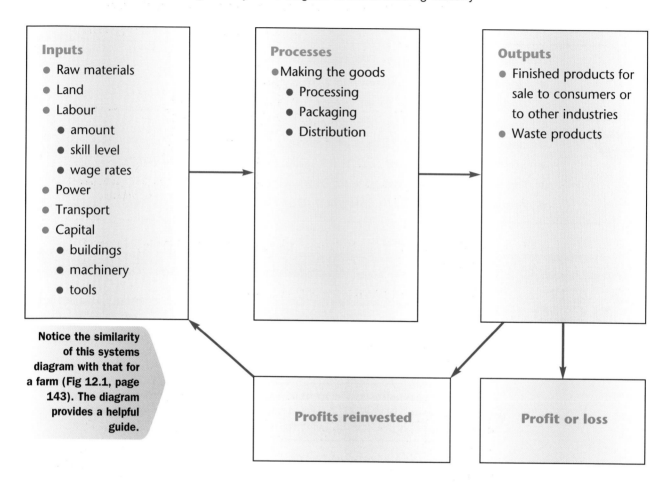

Inputs
- Raw materials
- Land
- Labour
  - amount
  - skill level
  - wage rates
- Power
- Transport
- Capital
  - buildings
  - machinery
  - tools

Processes
- Making the goods
  - Processing
  - Packaging
  - Distribution

Outputs
- Finished products for sale to consumers or to other industries
- Waste products

Profits reinvested

Profit or loss

Notice the similarity of this systems diagram with that for a farm (Fig 12.1, page 143). The diagram provides a helpful guide.

Manufacturing systems are called 'open systems' because only part of the income from sales is used for inputs and reinvestment. Some is taken by the owners or shareholders as profit.

PROGRESS CHECK

1. Divide the following occupations into primary, secondary and tertiary industry:

Coal miner	Dairy farmer	Taxi driver	Doctor
Comedian	Teacher	Bricklayer	Garden designer
Tax collector	Cabinetmaker	Hairdresser	Laboratory technician
Waiter	Journalist	Shopkeeper	Factory worker

2. For an industry or factory you have studied, draw a systems diagram.
3. Make a list of at least three occupations found in your local area for each of the four types of industry. Present your list as a table.

1.
P	P	T	T
T	S	T	T
T	S	T	S
T	T	T	S

Factors affecting the location of industry

AQA A   AQA B
EDEXCEL A
OCR A   OCR B   OCR C
WJEC B
NICCEA

KEY POINT Secondary industry or manufacturing industry involves the processing of raw materials or other manufactured products to produce goods.

Fig 13.2 Spider diagram of the factors affecting the location of industry

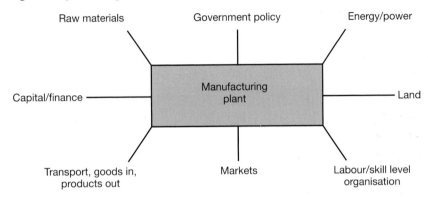

Industrial location factors for manufacturing industry

Fig 13.3 Pull factors on the location of a factory

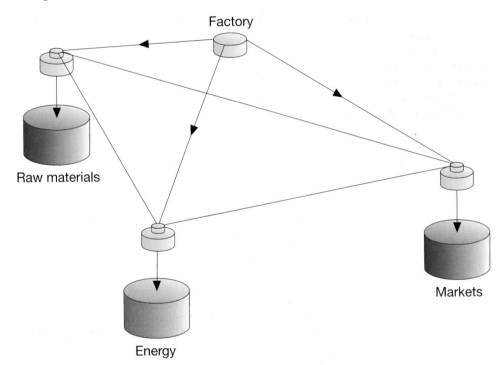

Raw materials

These are essential for the production of goods. In the past they have been very important in the location of industry. Some raw materials are very bulky or heavy and were expensive to transport. As a result, factories tended to be located near to the raw material. This was true of the iron and steel industry where the raw materials, iron ore and coal, were bulky and used in large quantities. Costs were reduced if the iron and steel plant located near to iron ore and coal deposits. The steel produced, though heavy, was less bulky and easier to transport than the raw materials. The iron and steel industry in the Ruhr, in Germany, was a good example of this factor. Another example would be the manufacture of cement in the Peak District of Derbyshire. Several tons of limestone is used to manufacture one ton of cement. As a result the manufacturing plant is next to the limestone quarry. These industries can be described as 'weight losing'.

Some products **gain weight or bulk** during manufacture. Baking, brewing and soft drinks are typical of this type. A bag of flour is compact and produces a much greater volume of bread. Beer and soft drinks contain large quantities of water. These 'weight-gaining' industries used to locate near to their markets to reduce transport costs.

Energy or power

Industries need **power** to operate their machinery. In the past industries relied upon steam power and tended to be located near to coalfields. Today they use **electricity** that is generated in power stations and transmitted over long distances using the **National Electricity Grid**. This has released most industry from the need to be close to a source of power.

The manufacture of aluminium takes very large amounts of power. As a result some aluminium manufacturing plants have located close to sources of hydro-electricity power to reduce their energy costs.

Land

Larger industrial plants, such as the steel industry, prefer flat land. This makes it easier and less costly to build and operate the plant. It is also helpful if the site provides space for expansion in the future.

Capital or finance

To set up, run and improve an industry takes large amounts of money. Even a small factory will need some capital to get started. All types of industry have to purchase raw materials, machines, premises and transport equipment. Banks, finance houses and governments offer loans to people (**entrepreneurs**) developing a business. These financial investors demand to see a plan of how things will develop and expect some return for the money they have invested. In some cases the source of capital may come from abroad.

Labour

The **availability**, **skill level**, **cost** and **working practices** of the workforce are all important factors in the location of industry. For example, the textile industry needs large numbers of workers who need not be highly skilled. As a result textile factories tend to locate near to or in large urban areas. In the nineteenth and first half of the twentieth centuries, Lancashire and Yorkshire had many textile manufacturing factories. Most of these have closed, as cheaper labour has now become available in South-East Asia.

Other industries need pools of educated and skilled labour that can be easily trained for new technologies. A good example is the computer hardware and software industries that originated in Silicon Valley in California but have now moved to the newly industrialised countries (NICs) like South Korea and Taiwan.

Productivity and worker attitudes are also important to employers looking to establish new factories. An area that is known for its stoppages, strikes and poor workmanship will not attract employers.

Factories need to be managed. When a site for a factory is chosen, the views of the owners and managers are important. They will have a view about the environment in which they wish to live and work. They may consider the quality of education available as well as leisure facilities such as golf courses.

Transport

Transport is needed to bring raw materials or the products of other industries to the factory and to deliver finished products to the market or other factories. A good transport network will help move products quickly. A site that is well connected to a **motorway network** with rail and airport connections will be more attractive. Sites in more remote or peripheral regions will have higher transport costs.

Markets

How have changes in one of these factors changed the location of industry, e.g. improvements in transport?

The goods produced are sold to meet the demand of the market. Sometimes people in the market will order goods from a factory. At other times, salesmen have to go into the market to persuade people to buy the product.

Industries that produce perishable goods will tend to locate near to the market. However, developments such as refrigeration have made this less necessary.

The people living in an area or country provide a domestic market. If they are affluent, their spending will encourage the growth of industry. In poorer LEDCs there is only a small domestic market and this hinders economic growth.

Where a factory sells nearly all its product to a small group of factories it is usual for the factories to group together. This is said to create an agglomeration and is an example of **industrial linkage**.

Government policy

There are a number of ways in which governments may influence the location of industry including:
- setting aside plots of land for industrial estates
- banning industries that create pollution from siting near to residential areas
- offering loans, subsidies and tax exemptions to companies willing to open factories in regions in need of development
- providing stable government, without corruption, so that overseas investors are attracted to the country.

Industrial inertia

A study of an established factory or industry in an urban area might show little obvious reason for its location. This may be because the original factors that gave it an advantage on the site no longer apply. However, despite the disappearance of these advantages, the factory continues to be successful. This may be because of its reputation. When this happens it is said to be a case of industrial inertia. If the factory were to be located today, another site with up to date advantages would almost certainly be chosen.

The special case of the footloose industries

Many of these factors remain important in understanding the location of industry. However a growing number of industries are free to locate in many locations. These are called the footloose industries. They include many of the growing industries.

PROGRESS CHECK

1. List the eight factors to consider when studying the location of industry.
2. What factors have influenced the growth between Slough and Reading in the UK?
3. How can governments influence the location of industry?

1. See pages 155–158 2. High-tech industries drawn to the M4, university towns, Heathrow Airport, south-east England 3. See page 157

Economic activity in MEDCs and LEDCs

AQA A AQA B
EDEXCEL A EDEXCEL B
OCR A OCR B OCR C
WJEC B
NICCEA

There are marked differences between the economic development of MEDCs and LEDCs. However for many LEDCs the gap is shrinking as transport and communications make links across the world easier.

There are also different levels of economic development within countries. Southern Italy and South Wales contrast with their economically prosperous areas in northern Italy and south-east England respectively.

> **Japan, Brazil and Bangladesh are countries at different levels of economic development.**

Industry in LEDCs

Barriers to industrialisation

> **It may not be appropriate for LEDCs to try to copy industrialisation in MEDCs. They may be more successful with low-tech, labour-intensive industry.**

Many LEDCs wish to increase their industries because they see it as a way to improve the quality of life of the people living there. However, there are barriers to industrialisation including:

- a lack of capital to invest and reluctance by investors from MEDCs following the failure of earlier schemes
- poor transport networks that are vulnerable to the weather
- unreliable sources of power, especially electricity
- lack of skilled labour and management expertise
- limited local markets

● unstable and corrupt governments
● unfair trading by MEDCs including quotas on imports.

The role of the transnational corporations (TNCs)

Transnationals are sometimes called **multinational companies**. They are very large companies with offices and production sites all over the world. Familiar names include Ford and IBM. They employ millions of people and are able to switch production to the cheapest site they can find. The turnover of some of these companies is larger than the GNP of some countries.

TNCs tend to locate in LEDCs to take advantage of:
● low labour costs
● government encouragement such as lower taxes, freedom from tariffs and quotas
● modern communications that allow the headquarters, research and development parts of the company to remain in an MEDC
● access to a potentially growing market.

LEDCs welcome TNCs because they provide employment, pay some taxes, bring in foreign currency (called forex), train the local workforce, improve roads and provide some schools and health facilities.

However, TNCs bring disadvantages including:
● they may distort government decisions to their own advantage
● most of the profit is transferred back to the MEDC
● the jobs created are low in skills
● wages are low and working conditions are often not good
● they may cause environmental damage
● they have their own plans and may leave a country for a cheaper source.

> **Know why TNCs locate in LEDCs and what advantages and disadvantages they bring.**

Newly industrialised countries (NICs)

NICs are former LEDCs that have developed their economy on the basis of manufacturing industry. They include Singapore, Hong Kong, South Korea and Brazil. Because of their rapid and aggressive rate of growth, South Korea and Taiwan have been described as '**Tiger**' economies, although they have been less successful in recent years.

The governments of the NICs have played an important part in developing their industry. Strategies they have used include:
● cheap loans and subsidies to new industries
● restriction on imports to protect their new industries from competition
● devaluation of their currencies to make exports cheaper
● tax holidays for overseas investors
● Free Trade Zones where no customs duties are paid
● strikes and other disruptive activity by workers are banned
● health and safety regulations for factories are not enforced
● priority on education and training to attract overseas companies.

> **It will be helpful to know a case study of one NIC.**

Industry in MEDCs

The iron and steel industry in Japan

Japan is one of the leading steel producers in the world. It exports its steel to the USA, South Korea, Malaysia, Singapore and China. The industry began in 1901 but most of the plant was destroyed after the Second World War in 1946. In the 1950s it grew again with modern equipment. The two main production areas are on the **Kanto Plain** between Tokyo and Yokohama and the Nobi Plain between Kobe and Osaka. The main characteristics of the industry are summarised here.

> **Case studies earn extra marks, especially if you can add the names of places.**

- Japan has little flat land and concentrates its steelworks on small coastal plains.
- A coastal location allows land to be reclaimed from the sea for expansion.
- Most of the raw materials are imported from Australia and the USA.
- Coastal locations help with the import of raw materials and export of finished steel.
- Coal-burning power stations are located near to the steelworks.
- Finance is available from banks in this wealthy country.
- Most Japanese live on the coastal plains and provide labour and a market for the steel industry.
- Transport facilities including deep-water ports are available close to the steelworks.
- There are many industries nearby, including motor car manufacture, that need steel.
- The government is encouraging the development of new steelworks away from the crowded coastal plain.
- Because of competition, particularly from China, the Japanese have been producing more specialised steel in recent years.

Fig 13.4 The inputs, processes and outputs of the steel industry in Japan

Inputs	Processes	Outputs
• Raw materials including coal, iron ore, limestone, scrap iron, oxygen, water, manganese, chromium, tungsten and vanadium • Land, using the coastal plains and land reclaimed from the sea • Labour, including the large, educated and skilled population living on the coastal plain • Power from coastal coal- and oil-burning power stations • Transport from deep-water ports to countries around the Pacific • Capital is available from the financial markets in Tokyo	Coal, iron ore and limestone combine in a blast furnace, at temperatures of over 1000°C, producing pig iron. Pig iron and scrap metal are converted in four hours to steel in a basic oxygen or electric arc furnace. Rolling of steel ingots into sheets. Processes are integrated on one site. (See Fig 13.5)	• Pipes and tubes • Steel plate • Rolled steel sheet • Bars • Wire rods • Beams

Fig 13.5 Main processes in an integrated iron and steelworks

Coal   Iron ore   Limestone

Blast furnace

Molten iron

Scrap metal   Pig   iron

Electric arc furnace   Basic oxygen furnace

Molten steel

Continuous casting

Slabs   Billets

Pipes and tubes
Steelplate
Rolled sheet steel
Bars
Wire rods
Beams

Rolling

Fig 13.6 Industrial areas in Japan

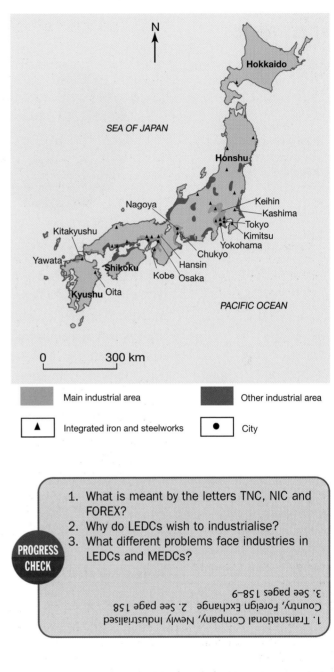

N

SEA OF JAPAN

Hokkaido

Honshu

Nagoya   Keihin
        Kashima
Kitakyushu   Tokyo
        Kimitsu
Yawata   Yokohama
        Chukyo
        Hansin
Shikoku   Kobe Osaka
Kyushu Oita   PACIFIC OCEAN

0   300 km

Main industrial area   Other industrial area

▲ Integrated iron and steelworks   ● City

PROGRESS CHECK

1. What is meant by the letters TNC, NIC and FOREX?
2. Why do LEDCs wish to industrialise?
3. What different problems face industries in LEDCs and MEDCs?

1. Transnational Company, Newly Industrialised Country, Foreign Exchange 2. See page 158 3. See pages 158–9

The changing structure of industry

AQA A
OCR A OCR C
WJEC A
NICCEA

Fig 13.7 The structure of indusry shown on a pie chart and a triangular graph

 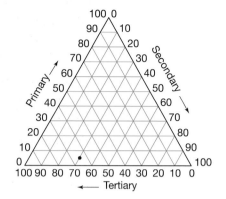

> Pie charts and triangular graphs can be used to show the structure of industry.

The **structure** of the industry of an area can be described using the percentages of workers in primary, secondary and tertiary employment. Many LEDCs tend to have a large proportion of their economically active population engaged in primary economic activity. When **industrialisation** takes place, employment in manufacturing grows and fewer people have jobs in primary economic activities. This has happened in the Newly Industrialised Countries (NICs) such as South Korea, Taiwan and Singapore. Many MEDCs, such as the UK, USA and Germany have moved to a stage of **de-industrialisation** where their employment structure shifts away from manufacturing towards service industries.

New sites for industry

AQA A AQA B
EDEXCEL A EDEXCEL B
OCR A OCR C
WJEC A
NICCEA

> **KEY POINT** Governments and local authorities can take action to encourage industrial growth.

Science and business parks

Many local authorities have tried to attract new industrial employment by providing sites for science or business parks. These can be found on **greenfield or brownfield sites**. On these sites the local authorities, often with the aid of government or EU funding, provide roads and services as well as attractive landscaping.

Business parks are intended to attract offices and high-tech industries. They may also have some light industry or distribution services.

> You should know at least one example of each of these types of site, as they are very popular with question setters.

Science parks, such as the one to the north of Cambridge on a greenfield site, or that on the edge of the Aston University campus on a brownfield site in Birmingham, are for new high-tech, science and technology-based companies. They make use of the highly skilled research workforce of the nearby university. Companies act as partners with university researchers to develop a commercial product.

Development areas and growth poles

There are a number of reasons why some governments recognise the need to develop industry including:

- the regeneration of an area that has seen the decline of its industry and increased unemployment, e.g. the decline of iron and steel and pottery in Stoke on Trent, the closure of coal mines in South Wales, the reduction of shipbuilding on Tyneside and in Belfast
- to encourage the setting up of industry in an area for the first time, e.g. growth poles in Brazil to attract people away from the crowded areas around Rio de Janeiro and São Paulo or rural development grants in remoter parts of Wales and Scotland.

Governments can encourage industry by establishing a **development area** where new companies can obtain benefits including:

- grants for the establishment of new factories
- ready-built premises with low rents
- small premises to allow the growth of new companies
- good road links to motorways and railway container depots
- advertising to promote the advantages of the area.

Enterprise Zones

These are special zones within areas of high unemployment and industrial decline where taxes are reduced and planning restrictions relaxed for ten years.

A case study of London Docklands would be useful here.

Urban Development Corporation

Following the success of Development Corporations in the building of new towns, the same strategy has been adopted for the redevelopment of some urban industrial areas. In this strategy, a piece of land is taken away from local government and a Development Corporation established. The Corporation has the responsibility, with the help of government and private finance, of redeveloping the area. This strategy was used to redevelop London Docklands. This derelict site of redundant docks has been transformed into a centre of employment with new office blocks and housing. Although this has created employment much of it has been unsuitable for the people who were living in the area.

Despite their success and popularity, government policy is to restrict further developments of these centres because they damage town centres, encourage car use that is environmentally damaging and use greenfield land.

The changing location of the retail industry

The retail industry includes shops and was traditionally located in the **Central Business District (CBD)**. In addition, shops were found in suburban high streets and local centres. Since the 1980s the UK has followed the pattern set in the United States with the retail industry moving to the edge of the urban area. The types of **out-of-town** shopping centre to be seen include:

- superstores containing a large supermarket or hypermarket and frequently a petrol filling station and car wash

- retail parks for large single-storey outlets for garden supplies, furniture, carpets, toys, pet food and accessories, DIY stores, sports goods and clothing
- regional shopping centres which consist of an undercover shopping mall, with two or three major chain stores acting as anchors and up to 200 other shops, cinema screens, restaurants and huge car parks. Some of these centres are served by their own light railway
- outlet villages such those at Bicester in Oxfordshire, Colne in Lancashire and the one occupying the former railway works at Swindon where chain stores sell discounted lines, often from the previous season's fashions.

These centres have been a great success. They provide plenty of car parking, good access most of the time, a range of shops often in warm and dry malls, safety as they are patrolled by security guards and competitive pricing. They do cause some problems at peak times when tailbacks can block surrounding roads and motorway slip roads.

The development of **out-of-town** centres has caused problems for town centres including:

- the loss of large shops has made the centre less attractive to shoppers
- smaller local shops have closed as fewer shoppers visit the centre
- vacant premises are often filled with less attractive discount and charity shops
- out-of-town centres are less accessible to the elderly and those without cars.

The decline of manufacturing industry in MEDCs

AQA A   AQA B
EDEXCEL A   EDEXCEL B
OCR A   OCR B   OCR C
WJEC B
NICCEA

Traditional heavy industries such as iron and steel making have declined in many MEDCs. This is because:

- raw materials have become exhausted, e.g. iron ore in South Wales
- markets have moved to other countries, e.g. shipbuilding from Clydeside to the Far East
- competition from countries with cheaper labour, e.g. textiles from Lancashire to India and China
- older companies did not keep up to date by investing in new technology, e.g. car manufacture in the West Midlands.

The closure of **traditional industries** can have devastating effects upon an area. Unemployment spreads from the plant that closes to affect its suppliers. People in the area have less to spend so the local retail and entertainment industries suffer. People may not be able to continue their mortgage payments and house prices fall.

This was the case in the valleys of South Wales as coalmines, steelworks and associated industries closed or cut back.

PROGRESS CHECK

1. Describe how a triangular graph can be used to show the three sectors of industry for several countries.
2. Define each of the following:
   (a) greenfield site   (b) brownfield site   (c) science park   (d) business park.
3. Describe how the location of the retail industry has changed in recent years.

1. See page 162   2. (a) Greenfield site – land which has not been built on before; (b) Brown-field site – land which has previously been built on and which is redeveloped; (c) and (d) – See page 162   3. See page 163–4

LEARNING SUMMARY

After studying this section you should be able to:

- *describe different levels of development*
- *know the five stages of development*
- *understand plans to aid development*
- *describe the main patterns of world trade*
- *know the different types of aid*
- *understand strategies to achieve sustainability*

Every country strives for development so that it can improve its economy and raise the quality of life of the people living there.

> **KEY POINT**
>
> Development is the use of resources and technology to create wealth with which to improve the quality of life.

Different types of development

AQA A AQA B
OCR A OCR B OCR C
WJEC A WJEC B
NICCEA

Here are three examples of development from three countries.

Development in Brazil

In places the Amazon rainforest has been cleared for housing, agriculture and the Trans-Amazonian Highway as the country develops its timber and mineral ores. This development is raising the quality of lives of the people living there.

Development in India

The Green Revolution has been applied to agriculture in the Punjab. Income and living standards have improved for many people.

Development in the Caribbean

Tourism has been introduced to replace declining sugar cane and banana industries. Beach resorts, roads and power supplies have been provided and the jobs available are improving people's lives.

What is development?

 AQA A   AQA C

OCR A OCR B   OCR C

WJEC A WJEC B

NICCEA

Development is the process of change which improves the well-being of a society in terms of material wealth and quality of life. Development includes:

- better food supply
- decreased infant mortality
- secure employment
- access to education
- warm dry housing
- improved health care
- longer life expectancy
- better working conditions
- security in old age
- water supply and sanitation.

> Be able to list these 'quality of life' features. If you are asked how the quality of life in an area can be improved you can use these as headings.

There are four kinds of development:

- **economic development** including greater income and wealth through industrial growth
- **social development** with better standards of living, access to education, health, housing and leisure
- **environmental development** bringing improvements and restoration of the natural environment
- **political development** with progress to effective representative

Stages in development

 AQA C

OCR A   OCR C

NICCEA

Walt Rostow suggested a series of five stages of economic change through which a country passes as it develops.

Fig 14.1a Rostow's model of economic growth

Stage 1	LEDC Somalia	Traditional society	Subsistence agriculture, low technology, little capital and poor transport system
Stage 2	LEDC Zimbabwe	Beginning secondary industry	Development of primary industry products for export, transport network improves and secondary industry begins
Stage 3	NIC Taiwan	Economic take-off	Rapid growth of secondary industry, increased investment in agriculture, transport and services. Core area appears and urban population grows. Improvements in standard of living occur
Stage 4	MEDC Hungary	Drive to maturity	Development spreads from the core to all areas, complex transport systems are built, wide range of industry, rapid urbanisation and the decline of primary industry. Standards of living continue to improve
Stage 5	MEDC UK, USA	High mass consumption	High level of production and consumption. High standards of living

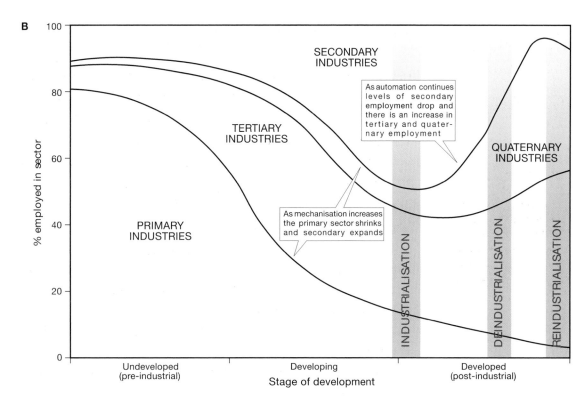

Fig 14.1b As a country passes through the five stages of development the proportions of primary, secondary and tertiary industry change

Measuring the level of development

There are number of measures of a country's level of development including:
- the changing percentage of labour in primary, secondary and tertiary jobs
- birth and death rates and rate of population growth
- infant mortality and life expectancy at birth
- adult literacy (being able to read and write) and access to secondary education
- **Gross National Product (GNP)**
- health measures such as people per doctor and access to clean water
- **Human Development Index**, created by the United Nations in 1990 to provide international comparisons. The index varies from 0 to 1 (most developed), and includes life expectancy, literacy, years in education and income per person.

> **KEY POINT**
> **Gross National Product is the total value of all goods and services produced by a country in one year, divided by the population total to give an average amount per head.**

> You may be presented with a table of data about different levels of development. Read the data and units used carefully. If you have to make comparisons they can be between rows or columns.

Attempts to measure development have been criticised. For example, GNP does not include subsistence production for the person's own use. In addition, national figures may not be accurate and could be distorted by corrupt governments. Within a country there will be variations from the average national figure, e.g. between northern and southern Italy, south-east Brazil and Amazonia, and between cities and rural areas.

Table 14.1 Applying the measures to selected countries

Country	Birth rate per 1000	Death rate per 1000	Life expectancy (years)	Adult literacy rate (%)	People per doctor	Trade balance (million US$)	GNP per head (US$)
Ethiopia	48	18	47	33	38 000	–683	100
Brazil	25	7	66	82	729	7 561	3 020
Japan	10	8	79	99	613	121 825	31 450
UK	13	11	77	99	623	22 183	19 950

The main **development characteristics** of LEDCs are:

- birth rates are high and death rates are falling which leads to population increase and a strain on resources
- high infant mortality rates because of poor health services, inadequate diet and poor housing mean people have more children in the hope that some will survive
- lack of protection against killer diseases such as malaria and AIDS
- lack of finance to provide enough homes and remove some of the squatter camps/shanty towns
- increases in life expectancy increase the non-economically active population and puts a further strain on resources
- low levels of literacy, as education is not available (especially for females) or is expensive, leads to unskilled workforce
- lack of food or poor quality food leads to low work output
- shortage of clean water allows spread of diseases such as cholera, dysentery and diarrhoea.

Influences on development

There are various factors involved in a country's **development**:

- political influences including political system, colonial influences and friendly nations
- physical environment including rainfall and water supply, natural hazards
- economic influences such as industrial structure, mineral wealth, agricultural potential.

Solutions to the problems of development

The solutions to the problems of development include:

- trade and aid
- sustainable development
- development projects in LEDCs
- appropriate technology
- role of the TNCs
- regional development plans.

PROGRESS CHECK

1. How could you measure the difference in development between an LEDC and an MEDC?
2. Describe the main features in the five stages of the Rostow model.
3. Identify some solutions to the problems of underdevelopment in LEDCs.

1. See page 167 2. See page 166 3. See page 168

Trade and aid

 AQA A AQA C
OCR B OCR C
WJEC A WJEC B
NICCEA

Fig 14.2 The North–South development divide

Fig 14.2 shows a map of the world divided into two areas. The map was drawn up by an international committee chaired by the former German Chancellor, **Willy Brandt** in 1980. To the north of the line are the developed richer MEDCs while to the south are the 'Third World' LEDCs. Since 1980 the gap between many MEDCs and LEDCs has become larger although some LEDCs such as Brazil, Malaysia, Indonesia and South Korea have developed considerably.

Trade

 AQA A AQA C
OCR B OCR C
WJEC A WJEC B
NICCEA

KEY POINT

Imports are goods purchased from abroad and brought into a country. Exports are goods purchased by other countries and sent to them. The balance of trade is the difference between the money earned from exports and that spent on imports.

Fig 14.3 Trade relationships between MEDCs and LEDCs

Exports: manufactured and processed goods, expertise
e.g. machinery, locomotives, aircraft, ships, bridges, generators

| MEDCs | Trade encouraged by GATT | LEDCs |

Trade restricted by tariffs, quotas and subsidies to protect home industry

Imports: primary products (raw materials and food)
e.g. sugar, tobacco, spices

Pattern of trade

Trade is the movement of goods and services between producers and consumers. Governments and companies aim to purchase the cheapest goods available for a given quality. However, their choice also depends on trade agreements and other political factors. Japan, the EU and USA account for 60% of the world's trade.

Fig 14.4 The pattern of world trade

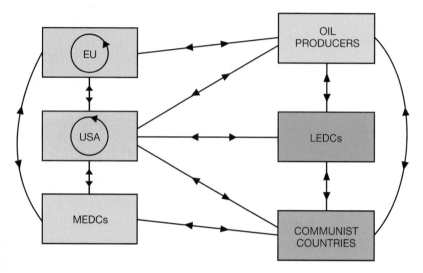

Trading agreements

> A useful class discussion on how to make trade more fair would be worthwhile.

Groups of countries make agreements to encourage trade between themselves. Members of the group receive better prices and conditions on trade that takes place within the group. **Trading agreements** include:

● the General Agreement on Trade and Tariffs (GATT) agreed in 1993 between 105 countries to increase international trade
● the European Union (EU) has an agreement between over fifteen European countries to make trade between themselves easier
● the North American Free Trade Association (NAFTA) which includes Canada, Mexico and the USA, and the Association of South-East Asian Nations (ASEAN) which includes Singapore, Malaysia and Indonesia, are free trade associations
● the Andean Pact includes Bolivia, Columbia, Ecuador, Peru and Venezuela
● the Organisation of Petroleum Exporting Countries (OPEC) tries to agree world price and output of oil
● the World Trade Organisation (WTO) encourages fair trade between countries whilst protecting workers against exploitation, child labour and trade barriers and promoting safe working conditions.

Aid

AQA A AQA B AQA C
OCR A OCR C
WJEC B
NICCEA

Why is aid needed?

Aid is the movement of resources from an MEDC to an LEDC and can include money, equipment, food, training, expertise or loans. The United Nations

Questions on aid are not popular with candidates. You should draw up a table of the advantages and disadvantages of the different types of aid for the giver and receiver.

encourages countries to spend 0.7% of their GNP on aid but few do. Aid is intended to help LEDCs to continue their development and improve the quality of life of the people there. Aid may be given to help with a sudden short-term problem or for a longer-term project.

What are the different types of aid?

- **Emergency aid (short-term)**, following a natural disaster such as an earthquake, volcanic eruption, severe storm or war, usually consists of food, medical supplies, tents, clothes and water.
- **Official bilateral/multi-lateral aid**, such as between the UK and Commonwealth countries or between the United Nations, International Monetary Fund, World Bank and an LEDC, exists to fund large projects but has the downside of debt repayments.
- **Charitable, voluntary or non-governmental aid**, from charities such as OXFAM, Christian Aid, Band Aid, Action Aid, gain income from fund-raising events and provide gifts to fund relatively small-scale projects.
- **Conditional (tied) aid**, such as the cancelling of debt in return for better financial management or agreement to buy particular products from an LEDC, is given to help a country's development.
- **Long-term aid**, such as education and training for farmers and young people, is designed to produce longer-term and lasting improvements in the quality of life and infrastructure through development projects.

Appropriate technology and sustainable development

KEY POINT | Appropriate technology and sustainable development are alternative ways to reduce the gap between rich and poor.

Appropriate technology

Some aid has been criticised as being too high-tech for LEDCs. The local people lack the expertise to maintain the project and have difficulty sustaining it. Spare parts are expensive or not available in the LEDC. Appropriate technology projects are low-tech, cheaper, use local materials and there is less to go wrong. They leave local people in control and there is less bureaucracy. Appropriate technology schemes often make use of the abundant local labour supply; products are cheap and local people can afford them.

E.F. Schumacher used the phrase 'small is beautiful' in 1973 for this type of development.

Examples of appropriate technology projects include:

- installing simple bamboo water pumps in villages
- providing bicycles for farmers to get goods to market
- using local streams to generate small amounts of hydro-electricity
- collecting rainwater in large clay pots
- using cement and chicken wire to reinforce walls and roofs.

Sustainable development

This refers to development which not only meets people's needs today but also those of future generations. We are moving towards **unsustainable development** as people consume more and more energy in using their mobile phones, the Internet, cars, air travel and air conditioning. If population and energy consumption continue to increase, the future of resources and the environment becomes more uncertain.

In 1992 the United Nations produced Agenda 21 with the following key points.

- Sustainable development is a local and a global concern involving local people and national governments.
- Local people and national governments are stewards of the Earth and should study how they consume resources and plan for a sustainable future.

The seven **key concepts** of sustainable development are:

> The slogan for Agenda 21 is 'think globally and act locally'.

- **Interdependence**: people, the economy and the environment are interdependent at a range of scales from the local to global.
- **Citizenship and stewardship**: people have rights and responsibilities including participation and co-operation.
- **The future**: responsibility extends to protecting the needs and rights of future generations.
- **Diversity**: recognition and respect for the cultural, social, economic and biological diversity of our environment.
- **Equity**: acknowledgement and respect for the quality of life, equity and justice for all people.
- **Sustainable change**: plan and monitor change to ensure sustainability of developments and carrying capacity.
- **Uncertainty**: understanding of the uncertainty of outcome and taking appropriate precautions.

Examples of sustainable development

Commercial fishing

Sea fish stocks are being used up faster than fish are breeding. The European Union has imposed quotas, regulations on net size, closed seasons and exclusion zones to conserve stock. This has proved a tough policy for politicians, fishermen and consumers alike. However, it is vital if fishing is to continue in the long term.

Forests

> Sustainable development is a key feature of every GCSE Geography syllabus. You must understand the term and know of examples in the UK and nationally.

Forestry provides a source of income and employment for some countries. It is very difficult to stop forest clearance and deny land to poor landless peasants who live near to forests, particularly tropical rainforests. Their governments may need the foreign exchange to improve agriculture or education and there is demand worldwide for their hardwood timber. Sustainable forestry allows some mature trees to be removed without decreasing biodiversity. The trees are cut and removed with care and new trees planted. This is hard to supervise in some of the more remote forests.

Building on brownfield land

The UK needs more houses and space for commerce and industry. Local authorities have been encouraged to develop derelict sites inside urban areas. Old factories, coal yards, railway sidings, gas works and military sites are being built on. This brownfield land may need clearing, access is not always easy and the local area not all that desirable. Developers and house buyers prefer housing on greenfield sites but building inside the city can regenerate it, cut travelling, reduce the need for new roads, preserve agricultural land and allow urban communities to be regenerated.

PROGRESS CHECK

1. Why do some countries form trading agreements with other countries?
2. Identify the different types of aid.
3. What do you understand by the terms 'appropriate technology' and 'sustainable development'?

1. See page 170 2. See page 171 3. See page 171

15 Tourism

LEARNING SUMMARY

After studying this section, you should be able to:

● define tourism and describe the ways in which it is changing
● describe the growth and characteristics of tourism in LEDCs and MEDCs
● understand the factors contributing to changes in tourism
● recognise the impacts that tourism is having on places

Leisure and tourism

AQA A   AQA B   AQA C
EDEXCEL A   EDEXCEL B
WJEC A

> Recreation becomes tourism when the person spends at least one night away from home.

Many people in MEDCs have free time after their day at work. After completing their duties at home they have time for leisure activities. In addition most people in MEDCs have paid holiday periods during the year. These trends are also appearing for people in paid employment in many LEDCs. People use this leisure time for recreation and tourism. Recreation is leisure time that lasts less than 24 hours. Recreation includes entertainment, skill improvement, rest and relaxation.

KEY POINT

> Tourism is a major earner and source of employment in the UK and many other countries.

What is tourism?

AQA A   AQA B   AQA C
EDEXCEL A   EDEXCEL B
WJEC A

Tourism is about the increasing voluntary movement of people, for a limited number of days (temporary), from their place of residence to visit another place. It involves the places they pass through, those who make their trip possible, and people living at their destination.

Tourism is a tertiary or service industry that provides relaxation and enjoyment for one group of people and is a source of income, through a wide range of jobs, for others. Some of the jobs provided will be directly related to the tourists, e.g. waiters and guides, while others, such as doctors and shopkeepers, are less directly involved.

People become tourists for a number of reasons including:

● recreation
● to play or attend a sporting event
● to visit friends
● for adventure
● to celebrate

● to meet health needs
● to experience foreign culture
● to escape daily routine
● for prestige or status
● for education.

> Remember that tourism involves an overnight stay and should not be confused with recreation.

Domestic tourists visit places in their own country while international tourists visit other countries.

The growth of tourism

Tourism is the **largest** and **fastest-growing industry** in the world today. It is estimated that 10% of the people in paid employment work in the tourist industry and that there are over 700 million tourist arrivals worldwide (Table 15.1).

Table 15.1 International tourist arrivals worldwide and by region

Region	1985 world share (%)	1999 world share (%)	1985-1999 annual growth (%)
World	100	100	5.5
Americas	20	20	5.6
Africa	3	4	7.5
Asia	10	15	10.0
Europe	65	59	4.5
Middle East	2	2	2.6

Many factors account for the growth of tourism in some places. They can be divided into two groups (Table 15.2).

Table 15.2 Demand and supply factors in tourism

Demand factors: Factors increasing the number of people wanting to be tourists	Supply factors: Factors encouraging the development of tourism
• People have increased amounts of income available for holidays (greater affluence) • Jobs provide paid holidays • The length of the working week has been reduced • Travel is easier • Cheaper package holidays are available • Advertising increases awareness • Hectic urban lifestyles • Modern telecommunications, including telephones, fax machines and the Internet make international booking more efficient • Early retirement and a longer life expectancy mean that more people over 60 are tourists	• Attractive climates, including warm to hot temperatures, little rain, or snow in winter for skiing • Culture including buildings, museums, art galleries and festivals, distinctive food, traditional music • The combination of sun, sea and sand • Safety and stable government • Special festivals or sporting events • Nostalgia or adventure • Scenery such as mountains, gorges, waterfalls, coasts and caves • Ecology including interesting trees, plants, birds and marine life • Famous sites of historic events • Sports facilities and nightlife

Types of tourist break

Be ready to account for the growth of tourism in some places but not in others.

Be ready to suggest how each of these could be applied to places you know.

There are many options available to tourists including:
- **package holidays** which include flight, accommodation, food and a guide
- **weekend breaks** at hotels mainly used by business people during the week
- **family holidays** designed to provide particularly for children with organised events, child minding facilities and evening entertainment
- **age-specific holidays** such as those for the over-55s or younger people between 18 and 30
- **self-catering** in an apartment, small cottage or house where holiday makers cook most of their own meals
- **fly-drive** which involve a flight and rental car and where hotels and itinerary may be arranged in advance
- **individual or independent group**, may be on a low budget or for an extended period such as backpacking across an area
- **specialist breaks** to develop a skill including sailing, painting, photography, gardening.

The distances tourists travel and the seasonality of their trips

Television travel programmes and newspaper advertisements can give you information about the popular places people visit.

Efficient and cheaper air travel allows people to travel all over the world as tourists. The numbers travelling tends to decrease with distance. In the UK, Europe remains the most popular destination, while in the USA, California and Florida are popular. Some people **enjoy returning** to familiar places and following the same routine year after year. Others search for **new experiences** in unfamiliar locations. These people travel increasing distances. The mountains of South America, the foothills of the Himalayas and Antarctica have become tourist goals.

Tourism can be a very **seasonal activity** with people looking for sunshine and coasts in the summer. In winter a significant number of tourists seek out snow for winter sports. This seasonality can cause problems for destination areas. People managing resorts attempt to extend their season, for example:

- seaside resorts use elaborate illuminations and put on carnival events to attract visitors out of season
- ski resorts have way-marked summer walking routes
- Mediterranean resorts offer long breaks for the over-50s during the cooler autumn and spring periods
- extensive indoor facilities, such as those at Center Parcs, provide resorts that are not dependent on the weather.

PROGRESS TEST

1. Suggest difficulties you might have in defining and measuring the growth of tourism.
2. Why does tourism develop in some places and not in others?
3. Why do both busy and quiet seasons pose problems for holiday resorts?

1. Not all travellers are tourists. Some people mix business and pleasure; no official records are kept. 2. See page 175 3. Pressure in busy seasons and unemployment and lack of income in quiet seasons.

National Parks

AQA A   AQA B   AQA C
EDEXCEL A   EDEXCEL B
OCR A   OCR B   OCR C
WJEC B

KEY POINT National Parks are attractive areas that are managed for the enjoyment of tourists and the benefit of local people.

There are National Parks in the USA (where the idea began), the UK, Malaysia and several African countries. A National Park is a large area of attractive countryside, often including small towns and villages, which is protected by law. In the USA, the National Parks tend to be wilderness areas.

National Parks in England and Wales

In 1949 the government passed the National Parks and Access to the Countryside Act to identify National Parks. In England and Wales there are ten National Parks of which the Peak District was the first. In addition the New Forest and South Downs are becoming National Parks and the Norfolk Broads is a Specially Protected Area.

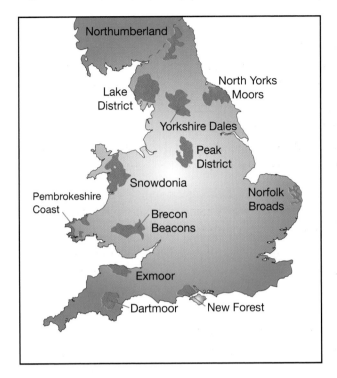

Fig 15.1 National Parks in England and Wales

Within **National Parks** the beauty of the landscape is preserved, the public is given access to much of the countryside; wildlife and important buildings are protected and traditional farming methods encouraged. The largest landowners are the National Trust, Forestry Commission and Ministry of Defence although well over 90% of the land is privately owned. A **National Parks Authority (NPA)** administers each National Park. Each Authority manages an area where economic activity continues, e.g. agriculture, forestry and quarrying, where local communities find their livelihood and large numbers of visitors arrive to walk, cycle, climb, sail and visit places of interest.

National Parks contain beautiful upland scenery with hills, moors, lakes, and forests and, in nearly all cases, spectacular coasts. The growth of the leisure industry together with mass car ownership means that more people are visiting the National Parks. These changes have brought both opportunities and problems for these areas.

Opportunities

Get to know a case study of one National Park and to name examples of places in it.

- The growing number of visitors bring money into the area.
- Visitors provide jobs and allow shops and other services to thrive.
- Roads and railways are maintained.
- Communities remain alive as people stay to work locally.
- A rich cultural life survives as audiences are supplemented by visitors on holiday.

Problems

- Traffic congestion occurs as poor local roads become crowded at peak times.
- Car parks fill up and grass verges are damaged by illegal parking.
- Footpaths are eroded by the large numbers of walkers.
- Bridleways become muddy with the increased number of mountain bikers and horse riders.
- Some settlements and sites become over-crowded to the point where their original attractiveness is threatened (these are called '**honeypot**' sites).
- House prices rise out of the reach of local young people as second home buyers move in from the large urban areas.
- Local people convert houses into holiday cottages and reduce the number available to local people.
- Farmers have their working land invaded by visitors causing damage to fences, crops and animals.

Managing the problems

Methods used to manage the problem in National Parks include:

- providing park-and-ride schemes on the edges of the sensitive areas
- reinforcing footpaths and providing alternative routes
- focusing demand on one or two honeypot sites and accepting they will be sacrificed for the sake of other areas
- ensuring that affordable new housing is built for local people
- demanding that quarrying is landscaped during and after use.

Tourism in LEDCs

| AQA A | AQA B | AQA C |
| EDEXCEL A | EDEXCEL B |
| OCR C |

KEY POINT Many LEDCs are attractive destinations for rich tourists but lack stable management.

The growth of tourism

LEDCs such as the Caribbean, China, Kenya, Zimbabwe, Thailand, Malaysia, Mexico and Egypt have well-established tourist industries. They have developed their tourist industry as a way of improving their economies and the quality of life for the people living there. The reasons for the growth of tourism in LEDCs include:

- government projects to build new airports, roads and hotels have been supported by investment from MEDCs
- costs are lower in LEDCs and they become cheap places for people to visit
- companies promoting tourism have been able to grow by advertising holidays in LEDCs
- traditional holiday resorts, such as the Mediterranean, have become crowded with noise, pollution and high costs
- people are looking for new experiences including the exotic
- good weather is more reliable in some LEDCs
- long-haul flights have become cheaper as larger aircraft carry more passengers directly to their destinations.

> Many LEDCs see tourism as a way of getting started on the path to development. They need to have stable government if they are to succeed in the long term.

For example, in Zimbabwe, tourists are able to fly to Harare where there are many good hotels. The weather is very predictable and good roads allow access to the countryside. The country has many interesting ancient sites such as the Great Zimbabwe ruins, the spectacular Victoria Falls, several impressive dams, such as that at Kariba, and interesting local traditions including sculpture. Costs are very low and high quality food plentiful. Recently political unrest has threatened this growth.

Tourism brings advantages and disadvantages to LEDCs

The **advantages or benefits** brought by income from tourism include:
- economic development can be promoted
- schools and hospitals can be built
- hotels provide jobs for people and a market for farmers' products
- employment is created in a wide range of industries including taxi driving, the making of gifts and souvenirs, travel guides and entertainment.

The disadvantages, problems, issues and conflicts include:

- planning controls are not strong in many LEDCs and uncontrolled development, especially of hotels, can make coastal areas and islands unattractive
- fishermen and other traditional activities can be displaced by coastal development
- many tourist activities are owned by foreign firms and much of their earnings leaves the LEDC
- farming may change to grow the types of food liked by tourists or food is imported to the disadvantage of local markets
- local culture may become modified to suit the daily shows for tourists
- crime, prostitution and drug use may increase
- tourists may have little respect for local traditions and religions
- the jobs created are casual and poorly paid
- valuable water resources are used for swimming pools and showers
- damage may be caused to the ecosystem through poaching or over-use of sites.

> Notice how words can be interchanged. 'Benefits' and 'advantages' are positive effects. Disadvantages can include 'issues', 'problems', and 'conflicts'. Question setters select the right word to guide you in answering the question.

PROGRESS CHECK

1. How many National Parks can you name in England and Wales?
2. List advantages and disadvantages of the growth of tourism for people living in National Parks.
3. Name some countries and the attractions they have for tourists.

1. See page 177 2. See page 177 3. France – Paris, wine, food, climate, coasts, countryside; China – Great Wall, Forbidden City, Terracotta Warriors

MEDC case study: tourism in the Mediterranean

The Mediterranean coast

The coastline and islands of the Mediterranean provide many tourist sites for people living in the cooler and less sunny parts of Europe including the UK, Scandinavia and Germany. The eastern coast of Spain, the French Riviera, and the southern coast of Greece, as well as islands such as Majorca, Cyprus and Crete, are the main tourist areas. The main holiday season is from April to October and during this period the population of the coastal area can double. This causes problems for local services such as water and sewerage whilst boosting employment in the hotel and other service industries.

The coast of the Mediterranean is some distance from the industrialised core of Europe. Before the growth of tourism the Mediterranean coast was part of the poorer periphery of Europe.

> It can be helpful to use the terms 'core' and 'periphery' when describing distributions. Be sure of their definitions.

KEY POINT

The core is the area of a country or region where economic activity is at its greatest. Population, industry and commerce are concentrated in the core. In contrast, the periphery is remote from the core and has low population density, high unemployment and out-migration.

There are both costs (disadvantages) and benefits from tourism for people living in the Mediterranean.

Costs of tourism

The costs include:

- pollution: coastal settlements and industrial sites, such as oil refineries and chemical plants, discharge storm water, sewage and waste into the almost landlocked and tideless sea
- poorly-planned building development including hotels, shops and apartments
- loss of farm land to building as the price of land rises
- congestion, despite the building of intrusive new roads and motorways, as both visitors and local people have access to cars
- increased noise from flights and night clubs as family holidays tend to take up 24 hours with their different activities
- the occupation of local shops and services with souvenir and gift shops, fast food restaurants and places of entertainment
- increased crime, alcoholism, drugs and bad behaviour as young people have 'a good time'.

Benefits of tourism

> **Why do people from the economic core of Europe choose to visit the Mediterranean for holidays?**

The benefits include:

- a variety of jobs are provided in areas where farming and fishing or unemployment used to be the norm
- local people have access to a higher standard of living as tourists spend money
- better roads, hospitals and shops are provided to meet the needs of visitors
- local people are able to use the tourist attractions such as bars and night-clubs
- the tourists bring new ideas and interests to the area
- funds are made available to conserve and restore old buildings.

LEDC case study: tourism in Singapore

Singapore is a small crowded island on the tip of the Malaysian peninsula. It is located very close to the Equator. Singapore is a highly organised country that built its initial wealth upon the trade passing through it as a port. With the growth of global tourism, people in Singapore have set out to make it a tourist attraction despite its small crowded site.

The number of tourists arriving in Singapore has risen rapidly since 1990. Visitors, on arriving in Singapore while having their passports checked by the immigration authorities, are asked to fill in a form. The form asks them to state the purpose of their visit. Most of the visitors are from South-East Asia, Japan and Taiwan. The main purpose of their visit is for holidays and the average length of stay is just under a week. Quite a large number of visitors are in transit.

Singapore's attractions as a tourist destination

The main attractions of Singapore that contribute to the tourist industry are:

- the equatorial climate with sunny mornings and warm evenings
- flora and fauna in the two nature reserves
- bird watching

- coastal activities including sea and sand on the offshore islands
- cultural events and the arts and crafts associated with the Indian, Malay and Chinese communities
- the National Museum, Parliament building and Kanji War memorial are popular sites for visitors
- the annual shopping sale, the international Dragon Boat racing and the Festival of the Arts.

Sustainable tourism

AQA A AQA C
EDEXCEL A EDEXCEL B
OCR A OCR C
WJEC A

> Sustainability is a very important word in new GCSE syllabuses. Know what it is and learn about sustainable tourism.

The growth of tourism threatens the very places the tourists visit. The erosion caused by people walking across a site and the moisture from their breath causes damage. The Taj Mahal, Stonehenge, the caves at Lascaux in the Dordogne, the Great Barrier Reef in Australia and the pyramids in Egypt are just five sites where access has been restricted. Summer and autumn visitors to quiet villages in the Cotswolds and Lake District find the peace they came to find destroyed by the large numbers of people. Even Sunday climbers find themselves queuing to use the chalk-marked holds as too many people arrive in Snowdonia, North Wales, for a mountain experience.

People managing tourist sites or planning new ones are looking at alternative ways in which to develop them. They want the tourism to bring the revenue but the attractions of the site to be sustained. Sustainable plans include:

- ecotourism
- honeypot sites and wilderness areas.

KEY POINT

> Ecotourists visit areas to learn about the people living there, the local culture and environment without causing damage. At the same time, ecotourists want the local people to benefit from the economic advantages that they bring.

Ecotourism

Ecotourism is taking place in some areas. It involves training local guides to take tourists to a variety of locations so that no one place is visited by too many people. During the visit, the guide makes sure that no damage is caused and the visitors are able to enjoy the wildlife and vegetation around them. It also means teaching local people to respect and protect their own local environment. Some of them may be tempted to make quick money by opening conventional tourist attractions or selling land to developers. They are shown that ecotourism brings increased income whilst protecting and improving their way of life.

The advantages of this approach are:

- some of the income is used to protect the natural environment
- the quality of local people's lives is improved without their culture and way of life being destroyed
- local people are able to remain on their land as farmers as well as working in the tourist industry
- the area is conserved for the benefit of future tourists.

No one is certain that this plan will work. Contact with visitors changes the ways in which local people want to live. Even small numbers of visitors lead to changes, as they demand quality hotels, fast communication routes, air conditioning, familiar food and sewerage facilities.

Honeypot sites and wilderness areas

This strategy involves identifying some popular locations and allowing them to develop as honeypot sites. Bigger car parks are built, more public toilets provided, hotels and restaurants extended, cycle rides and footpaths marked, access roads improved and housing built. These popular sites become even more popular and very crowded in the summer. Local residents may complain but the shopkeepers, hotel and restaurant owners thrive.

Allowing these honeypot sites to attract so many people draws demand away from other sites and they are protected. Around the less popular sites, access is not improved and car parking is very limited. These areas remain for farming or moorland and, in the more remote parts, as wilderness. In this strategy, some popular sites are sacrificed to protect all the others

PROGRESS CHECK

1. Describe the difference between the core and periphery of a country.
2. For a place you have studied, describe the costs and benefits it receives from its tourist industry.
3. What do you understand by ecotourism? How can it be developed in a place you have studied?

1. See page 179 2. See page 180 3. See page 181

Sample GCSE question

1. Study Fig 1.

Fig 1. Field sketch of a headland

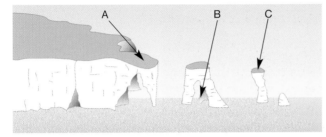

(a) Name the type of landforms labelled A, B and C. **[3]**

A = headland ✓, B = arch ✓, C = stack ✓. ← Learn these terms.

(b) Describe the **four** processes by which the sea erodes the coast. **[4]**

Hydraulic action is the force of the waves hitting the cliffs. This compresses the air into the cracks and joints, which weakens and fractures the rock into pieces ✓. The loose rock is thrown against the cliff by the waves and causes further erosion. This is called abrasion ✓. The loose material is gradually broken down into smaller and smaller pieces by attrition ✓. It becomes smaller and rounder, eventually becoming sand. Some rocks, such as chalk and limestone, are dissolved by the salt water; this is called corrosion ✓.

2. (a) Name **three** ways in which a coastline can be protected from erosion.

A coastline can be protected by hard defences, such as a seawall or groyne, or by a soft defence, such as beach nourishment. ← By selecting hard and soft defences your explanations in (b) are not so repetitive.

(b) Explain how each method works. Labelled diagrams will be credited. **[6]**

The seawall is built along the back of a beach and is made of thick concrete ✓. It is strong enough to stop the waves eroding the coast ✓ and prevent flooding.

A groyne is built across the beach and stops longshore drift from moving sand along the coast. More and more sand is caught by the groyne, which makes the beach higher and wider. The beach can now prevent even the biggest waves from eroding the coast ✓.

Beach nourishment is used if beach levels have become very low. Sand is dredged from offshore and dumped onto the beach, making it higher and wider. The beach is much stronger and able to stop all the waves ✓.

← Maximum of four marks if diagrams are not included. Labelled diagrams are a useful way to revise and can save time in the exam.

3. (a) Name and locate a coastal feature formed by deposition.

A beach has been deposited at Sennen Cove near Land's End ✓.

← You could have chosen a spit, a tombolo, a bar or a salt marsh. One mark for a clear location.

← Relate your answer to the place you have named. Maximum of three marks if you fail to do this.

(b) Describe and explain how it was formed. Labelled diagrams will be credited. **[7]**

Sennen Cove has formed in soft rock between two headlands of harder rocks ✓. The Cove is sheltered from the largest waves by Land's End ✓. The smaller waves are not able to carry all the sand and rocks, and deposit them in the sheltered Cove ✓. The beach is mostly made of sand ✓. The gentle waves make the sand into a wide, flat beach ✓. Below the cliffs, the beach is higher and made of large pebbles. These are left there by much bigger waves during stormy weather ✓.

← Maximum of five marks if diagrams omitted.

← Labelling your diagram to explain how the beach was formed can gain marks.

Sample GCSE question

1. Study Fig 1.

Fig 1 A constructive plate boundary

→ Direction of plate movement

(a) Explain why the plate boundary shown in Fig 1 is called a constructive boundary. **[3]**

> *It is called a constructive boundary because new crust is being made ✓. As plates move apart, magma is able to rise between them ✓. As the magma flows out across the ocean floor, it cools and forms new crust✓.*

(b) Name an example of a constructive boundary. **[1]**

> *The North Atlantic Ridge ✓.* ◄

| The East Pacific Rise is also correct. |

2. (a) Describe two types of volcanic activity that take place at the surface. **[4]**

> *Lava flows are rivers of molten rock ✓ which flow down the slopes of a volcano ✓. Huge quantities of ash ejected into the atmosphere ✓ fall on the surrounding area, burying crops and buildings ✓.*

| Two types of volcanic activity with a descriptive point for each. |

(b) Why do people live close to areas of volcanic activity? **[4]**

> *Weathered volcanic ash gives fertile soils ✓, which can be used for arable farming ✓. Tourists are attracted to areas of volcanic activity ✓. Local people can make a living providing accommodation, running gift shops and acting as guides for visitors wishing to visit the crater rim ✓.*

3. (a) For a volcanic eruption you have studied, explain the causes of the eruption. **[4]**

> *The eruption of the Mount St. Helens volcano in north-west USA was caused by magma rising from where an oceanic plate, the Juan de Fuca plate, moves under the North American continental plate ✓. As the plates grind together, friction causes great heat, which turns the oceanic plate into molten magma ✓. When sufficient pressure has built up over time ✓, it forces the magma up to erupt at the surface ✓.*

| The answer would benefit from a labelled drawing. |

| Maximum of two marks if answer is not related to an actual example. |

(b) Describe how the eruption affected people in the surrounding area. **[4]**

> *Sixty-three people died from the poisonous gasses that came from the eruption ✓. Large areas of forest were burnt and flattened together with some logger's camps. Loggers no longer had any employment and had to leave the area ✓. Landslides and mudflows blocked the North Fork River causing flooding. Local farmers suffered as cattle were drowned and crops washed away ✓. Over time, however, the thick layer of ash will improve the fertility of the local soils ✓.*

| Four points with evidence related to the case study used. Answer refers to short- and long-term effects. |

Sample GCSE question

1. (a) Give two sources of weather information used in weather forecasting. **[2]**

> *Satellites ✓, Stevenson's Screen ✓.* ◄——

(b) What instrument is used for recording air pressure? **[1]**

> *Barometer ✓.*

(c) What is the name of a line joining places of equal pressure? **[1]**

> *Isobar ✓.* ◄——

2 (a) Study Fig 1 and the weather forecast below.

Fig 1 Synoptic chart for 13 January 2001

'A cold, fine and bright day, with long, sunny spells. It is likely to be colder along the east coast, with even a chance of snow. Tonight temperatures will drop to −3°C and there will be a sharp frost.'

(a) What type of pressure system exists over the UK? **[1]**

> *A high pressure system or anticyclone exists over the UK ✓.*

(b) Explain the reasons for the weather described in the forecast? **[7]** ◄

> *Temperatures are likely to be low as it is winter in the Northern Hemisphere ✓. In January the UK receives low amounts of solar radiation, due to the low angle of the sun's rays ✓.*
> *Anticyclones are areas of sinking air. Skies will be cloudless giving sunny days ✓. At night the lack of cloud cover will let temperatures drop very low ✓, and since there is very little wind, frost is likely ✓.*
> *Cold air from central Europe picks up moisture as it passes over the North Sea ✓. The cold land cools the air, and there may be snow showers along the east coast ✓.*

3 For a place outside Europe which you have studied: ◄——

(a) Describe the main features of the climate.

> *Name the place and the type of climate found there. Descriptions should include:*
> - *maximum and minimum temperatures, temperature range and seasonality (summer/winter)* ◄——
> - *total rainfall, seasonality (wet and dry seasons; summer maximum) and reliability.*

(b) Describe how the climate affects the lives of people who live there. **[8]**

> *Answers may cover aspects of:* ◄——
> - *traditional and commercial agriculture (closely related to soils and vegetation)*
> - *exploitation of minerals* ◄——
> - *communications*
> - *tourism*

Sample GCSE question

1. Study Fig 1.

Fig 1

Age years	Country A	Country B
0–15	49%	20%
16–64	48%	61%
65+	3%	19%

(a) Compare the population age structure of the two countries. **[6]**

Country A has a greater percentage of the population in the younger age group from 0-15 with 49% ✓ compared with 20% in Country B ✓. Country A has fewer people in the 16-64 age range ✗. In the 65+ age group there is a higher percentage in country B ✓ (19% compared with 3%) ✓.

The command word is 'compare'.

Two clear comparisons with evidence are given. There is a mistake in the middle section since figures are percentages. Scores four out of six.

(b) Which of the countries is a developing country? Give two reasons. **[6]**

Country A is probably the developing country ✓ because it has such a large percentage of its population under 15 years ✓.
Developing countries have high birth rates because they do not have much birth control ✓ and large families can help on the farms ✓.
Also Country A has few people over 65 which means it has low life expectancy ✓ because of poor health facilities ✓.

Correct choice. Two reasons given with one of them more developed than the other.

This is a knowledge question where you have to recall case studies you have learned.

(c) Some countries are trying to increase their rate of population growth while others are attempting to reduce theirs. **[8]**

   (i) Name a country you have studied which is trying to either increase or decrease its rate of population growth.

The country I have chosen is China, which is trying to reduce its rate of population growth ✓.

Note that you have to choose one country.

A clear choice.

   (ii) Describe the ways in which it is trying to change its rate of population growth.

China is trying to reduce its population growth using its 'One-child policy' ✓. This means that couples are only supposed to have one child ✓. If they have one child they get help with housing and education ✓. China is also educating people, especially women, about birth control ✓. They have advertisements in the streets and on television about smaller families ✓.

Six marks out of eight, but there are other points that could have been made in describing the policy.

Sample GCSE question

1. Study Fig 1.

Fig 1

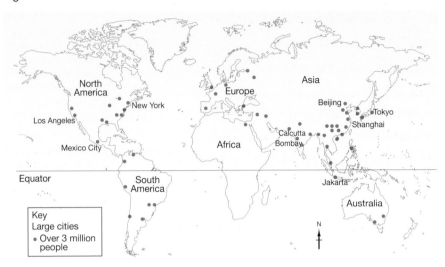

(a) Describe briefly the distribution of cities shown on the map. **[2]**

Most of the cities are north of the Equator ✓. There are most cities on the coast between Jakarta and Tokyo ✓.

(b) Some countries have only one large city. Suggest two reasons for this. **[2]**

These are called primate cities and are usually the capital city of the country ✓. They have become much bigger than all the other cities because they have a better harbour or resources like coal and iron ✓.

(c) Compare the advantages of using brownfield and greenfield land for new housing. **[8]**

Brownfield development is high on the government's agenda to solve the problems of housing shortages. The advantages of brownfield sites are that they already have some infrastructure ✓ and they are nearer to the amenities in the city ✓.

Developers like greenfield sites because they do not have to clean up the land before they use it ✓ and customers like living or working there ✓.

(d) For an urban area you have studied, describe and explain three ways in which it is trying to solve its transport problems. **[8]**

The urban area I have studied is Birmingham ✓. In Birmingham the transport problems have been solved by:

1. *Building an inner ring road with underpasses and flyovers like St Chad's Circus ✓. These help traffic to bypass difficult junctions ✓.*
2. *The central area of Birmingham is pedestrianized e.g. Corporation Street ✓ but buses are allowed into the area on special bus lanes. This encourages people to use public transport ✓.*
3. *They are building a new tramway system ✓ to bring commuters into Birmingham from the suburbs. It goes to Snow Hill station at the moment ✓.*

Sample GCSE question

1. Answer the following question with short answers.

(a) What do you understand by a tertiary industry? **[1]**

> *Tertiary industries are service industries like shops and banks ✓.*

(b) Name one raw material of the steel industry. **[1]**

> *The steel industry uses a large amount of iron ore ✓ that is turned into steel in blast furnaces.*

(c) In which sector of industry does a farmer work? **[1]**

> *Farmers work in primary industry ✓.*

(d) Name an industry where the product gains weight during manufacture. **[1]**

> *Beer gains weight during manufacture when water is added and that is why breweries used to be in cities to save on transport costs ✓.*

(e) Describe two ways in which governments can influence the location of industry. **[2]**

> *Governments can influence the location of industry by setting up industrial estates ✓ and by giving grants ✓. The government has just given Nissan a large grant to stay in Sunderland.*

(f) What do you understand by 'industrial inertia'? **[2]**

(g) Define a footloose industry, giving an example. **[2]**

> *A computer software company is footloose because it can be almost anywhere ✓ that the owner wants it to be but it will help if it is near a university ✓.*

2. For an LEDC you have studied, explain why the county is having difficulty in developing manufacturing industry. **[6]**

> *The LEDC I have studied is Nigeria. They are finding it difficult to develop industry because the people living there do not have the right skills ✓. Also the government has been unstable ✓.*

3. Name a Newly Industrialised Country (NIC) and describe its main characteristics. **[4]**

> *South Korea is a NIC ✓. i think it has developed a manufacturing industry with the help of its government ✓. They have provided education and training for the people living there ✓ and especially women who have joined the labour force ✓. i think the government has also put tariffs on imports to the country and this has protected the new industries ✓ in South Korea. There is a large labour force ✓ which i think has been willing to work for low wages ✓ and there have not been strong unions to have strikes to stop production ✓.*

When the examiner sets several short questions, be careful not to spend too much time on each. If your examination is two hours long you have about a minute per mark up to 100.

The candidate starts well with two short crisp answers.

In (c) and (d) the candidate forgets to put the question letter, this can be a problem for examiners.

A very good answer to (e).

The candidate cannot remember the answer for (f) so wisely leaves some space so that he/she can return to it later.

Country named correctly. The answer would gain two marks out of six. The reasons need to be expanded and linked to Nigeria.

A good choice and a splendid level of knowledge.

A number of candidates have adopted the habit of writing I as 'i' and sometimes adding a circle above the i. Do not do this.

Exam practice questions

1. Suggest **three** factors which influence the size and energy of a wave as it approaches the shore

   ....

   .... **[3]**

2. Study Fig 1

   Fig 1 Sketch of a coast

   (a) Describe the processes that have led to the formation of the spit.

   ....

   ....

   .... **[6]**

   (b) The houses in village A are threatened by the cliff eroding and retreating.

   (i) Describe the benefits of building a groyne at G...............................

   .... **[3]**

   (ii) Describe how the groyne might affect the spit. ....

   .... **[3]**

   (iii) Suggest an alternative method of preventing the erosion of the cliff, giving reasons for your choice.

   ....

   .... **[5]**

Exam practice questions

1. Study Fig 1. Describe the characteristics of the climate shown.

Fig 1 Climate graph for a location in the tropical rainforest of Singapore

    ....

    ....

    .... **[3]**

2. Describe the main characteristics of a tropical rainforest.

    ....

    .... **[5]**

3. 'Tropical rainforests are one of our most precious natural resources, yet they are being destroyed at a rapid and accelerating rate.'   *WWF-UK, 1998*

    (a) Give two reasons why 'tropical rainforests are one of our most precious natural resources'.

    .... **[2]**

    (b) With reference to an area you have studied, explain why large areas of the tropical rainforest are being destroyed.

    ....

    ....

    ....

    .... **[10]**

Exam practice questions

1. **(a)** Gross National Product (GNP) is one way of measuring development.
    Why is GNP not always a good measure of development?

    .... **[2]**

    (b) Suggest, giving a reason for each, two other different measures of
    development that could be used.

    ....

    .... **[4]**

    (c) Many MEDCs give aid to LEDCs. Describe one short-term and one
    long-term type of aid.

    ....

    .... **[4]**

    (d) Yelma, a small village in northern India, is connected to the main road via a
    ford. The village has electricity. The farmers in Yelma grow wheat and keep
    dairy cattle to feed their families; any surplus produce is sold at the market
    5km away in the local town. A development project, funded by an MEDC,
    has sent four experts to Yelma for two years to use local labour to:
    • build a road bridge to replace the ford
    • repair the dairy so that the milk can be processed and taken by lorry to
      the town
    • build a small agricultural training centre for village workers
    • restore the village meeting hall
    • install a village computer with Internet access.

    (i) Select, giving your reasons, which of these projects should be the
    first to go ahead.

    ....

    .... **[2]**

    (ii) Suggest two ways in which the success of the project could be judged.

    ....

    .... **[4]**

    (iii) What problems are likely to occur when aid from the MEDC stops after two
    years?

    ....

    .... **[4]**

Exam practice questions

1. (a) The use of coal for energy production in the UK has fallen in recent years. Identify two reasons for this.

    ....

    .... **[2]**

    (b) Suggest two reasons for the increased use of natural gas for energy production in the UK in recent years.

    ....

    .... **[2]**

    (c) Describe two advantages and two disadvantages of developing sources of hydroelectric power.

    ....

    .... **[4]**

    (d) 'Global warming could be worse in 100 years. An international group of scientists believes the temperature of the Earth's atmosphere could rise by 6°C in the next 100 years. With this change, sea levels could rise by 88cm. The scientists said their report should ring alarm bells in every national capital and every local community.' *The Guardian*, January 2001

    Outline the causes of these predicted changes in temperature and sea level.

    ....

    .... **[4]**

    (e) In an area that you have studied, there is a planning application to extract a resource, e.g. open cast mining, drilling for oil, the quarrying of limestone for cement.

    (i)   List four factors that should be considered.

    ....

    .... **[4]**

    (ii) Describe how two of the factors could be applied to your chosen area.

    ....

    .... **[4]**

Exam practice questions

1. **(a)** Identify four land uses that you would expect to find in the Central Business District (CBD) of a large city.

.... **[2]**

(b) Study Fig 1

Fig 1 The concentric zone model

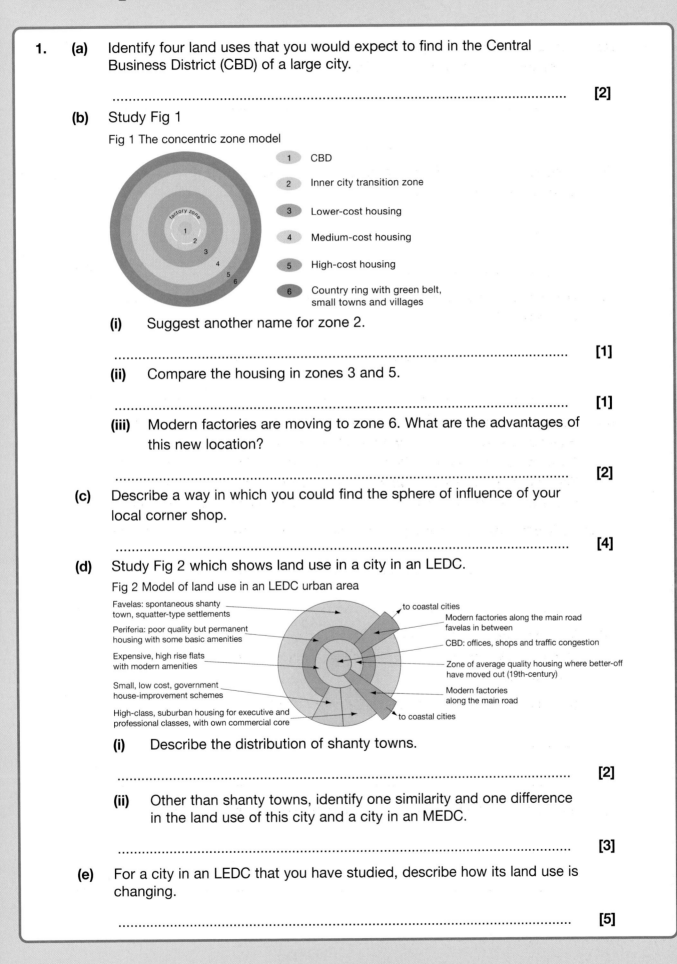

1 CBD

2 Inner city transition zone

3 Lower-cost housing

4 Medium-cost housing

5 High-cost housing

6 Country ring with green belt, small towns and villages

(i) Suggest another name for zone 2.

.... **[1]**

(ii) Compare the housing in zones 3 and 5.

.... **[1]**

(iii) Modern factories are moving to zone 6. What are the advantages of this new location?

.... **[2]**

(c) Describe a way in which you could find the sphere of influence of your local corner shop.

.... **[4]**

(d) Study Fig 2 which shows land use in a city in an LEDC.

Fig 2 Model of land use in an LEDC urban area

Favelas: spontaneous shanty town, squatter-type settlements

Periferia: poor quality but permanent housing with some basic amenities

Expensive, high rise flats with modern amenities

Small, low cost, government house-improvement schemes

High-class, suburban housing for executive and professional classes, with own commercial core

to coastal cities

Modern factories along the main road favelas in between

CBD: offices, shops and traffic congestion

Zone of average quality housing where better-off have moved out (19th-century)

Modern factories along the main road

to coastal cities

(i) Describe the distribution of shanty towns.

.... **[2]**

(ii) Other than shanty towns, identify one similarity and one difference in the land use of this city and a city in an MEDC.

.... **[3]**

(e) For a city in an LEDC that you have studied, describe how its land use is changing.

.... **[5]**

Exam practice answers

Exam practice questions: Coastal landscapes

1. Fetch, strength, depth of water offshore

2. (a) Processes leading to the formation of a spit:
   - waves erode cliffs and supply sand to the beach
   - prevailing winds and waves approach the shore at an angle
   - waves carry sand along the coast by the process of longshore drift
   - sand is deposited in the shallow water of the estuary where a river meets the coast
   - deposition continues across the estuary until tidal currents are strong enough to carry the sand away

   (b) (i) The groyne captures sand so the beach becomes higher and wider, absorbing more wave energy and preventing the waves from eroding the cliff.

   (ii) The spit is starved of a supply of sand so the waves have more energy and power to erode the spit, which becomes narrower. The sea may eventually break through the spit and destroy the salt marsh.

   (iii) Give a method (e.g. beach nourishment, rock armour, cliff stabilisation), describe the method and suggest advantages compared with the groyne, e.g. effect on the spit. Qualify your answer by recognising that your solution may also have disadvantages, e.g. cost.

Exam practice questions: Ecosystems

1. Temperature: hot all year; small range of temperature, 2°C
   Rainfall: high rainfall, over 2000mm annually; no month has less than 150mm

2. Main characteristics of a tropical rainforest:
   - very diverse number of species – up to 300 species of trees per km^2
   - trees are very tall (50m) and deciduous evergreen (shed leaves continuously)
   - forest has a series of layers; there is competition for light
   - shallow roots with buttresses to support height
   - large supply of dead leaves, branches etc. rapidly decompose in high temperatures and humidity – shallow roots tap rich nutrient supply
   - vast number of animal species adapted to live at various layers of the forest, e.g. monkeys in the canopy.

3. (a) Two reasons why tropical rainforests are one of our most precious natural resources:
   - biodiversity: many species are unique to the rainforest and a source of medical vaccines, e.g. anti-malaria
   - tropical rainforests help to maintain a stable atmosphere; forests absorb large amounts on carbon dioxide, an important greenhouse gas.

   (b) The Amazon rainforest would be a good case study. Reasons for destruction of large areas of tropical rainforest:
   - population growth
   - pressure on traditional agriculture
   - relocating urban poor
   - need for LEDCs to increase their GNP by
     – selling timber and forest products to MEDCs (for timber and paper pulp)
     – expanding agriculture (ranching)
     – developing hydroelectric power.

Exam practice questions: Development

1. (a) GNP does not include subsistence production. GNP is a national figure and does not reveal variations within the country. It may have been distorted by corrupt governments.

   (b) Birth and death rates; life expectancy; infant mortality; health measures such as access to doctors or hospitals; changing industrial structure from primary to tertiary industry; literacy.

(c) Short-term aid could be emergency aid following a disaster. Longer-term aid includes larger development projects such as building dams or setting up education and training programmes.

(d) (i) No one answer is the best and arguments can be put forward for all of them. The bridge is likely to be popular because it opens up the road to the market.

(ii) Increase in farmers' income; slow down of out-migration; local people healthier.

(iii) Lack of maintenance of bridge, dairy etc. means they are no longer used; trained people leave the village; crowding of village as people come in from countryside.

Exam practice questions: Energy

1. (a) Cost of extraction; high costs of transport; pollution; trade unions.

(b) Resources have become available; it is easier to transport and is a cleaner fuel than coal.

(c) Two advantages: clean, cheap, provides recreational lakes, sustainable
(2 marks at 1 each)
Two disadvantages: high set-up costs, floods farms and villages, attracts malaria, may silt up
(2 marks at 1 each)

(d) Greenhouse effect due to gases in the atmosphere. Gases increased by burning fossil fuels. Trapped solar radiation is increasing temperature.
(1 mark for each statement, credit specific detail)

(e) (i) • Increase in employment, transport, dust
• Hours of working
• Screening, restoration, drainage
• Alternative sources
(1 mark for each appropriate idea)

(ii) Names an area and applies two of the factors

Exam practice questions: Settlement

1. (a) Four different land uses could include commercial, retail, or offices and shops: restaurants, banks, department stores, government offices. Allow bus stations.
(Mark 2 correct as 1 mark, 4 correct as 2 marks)

(b) (i) Zone in transition, inner city, twilight zone
(1 mark)

(ii) Terraced housing with small gardens or yards; detached houses with gardens and car parking.
(One feature of each compared for 1 mark)

(iii) Greater space, less congestion, modern factories suited to up-to-date processes, closer to workers, closer to motorways.
(Two advantages 1 mark each)

(c) The best way would be to prepare a short questionnaire and interview the shop's customers. Reject any that are occasional users. Ask customers where they live. Ask enough customers to get a representative sample. Consider asking customers at different times of the day and week. Plot locations on a map of the area. Identify the sphere of influence on the map. You could draw a line around locations or lines from locations to centre of map (star map).

(d) (i) Shanty towns are located on the outskirts of the city, along some roads, with a small number by the river.
(2 marks at 1 each)

(ii) Similarities: both have a CBD and manufacturing areas along roads.
(1 mark)
Differences: the LEDC city has low-class housing around the edge and high-class near the centre. The opposite is true of an MEDC city.
(2 marks)

(e) Description of one change, with some place detail (2 marks)
Two changes with some place detail (4 marks)
Further change or more detail, e.g. redevelopment of the central area, new self-help housing, appearance of new shanty towns, enlarged bus station, construction of new roads and road improvements (5 marks)

Index

Index